Fractals and Chaos Simplified
for the Life Sciences

Fractals and Chaos Simplified for the Life Sciences

LARRY S. LIEBOVITCH
Center for Complex Systems
Florida Atlantic University
http://www.ccs.fau.edu/~liebovitch/larry.html

New York   Oxford
OXFORD UNIVERSITY PRESS
1998

Oxford University Press

Oxford New York
Athens Auckland Bangkok Bogota Bombay Buenos Aires
Calcutta Cape Town Dar es Salaam Delhi Florence Hong Kong
Istanbul Karachi Kuala Lumpur Madras Madrid Melbourne
Mexico City Nairobi Paris Singapore Taipei Tokyo Toronto Warsaw

and associated companies in
Berlin Ibadan

Copyright © 1998 by Oxford University Press, Inc.

Published by Oxford University Press, Inc.,
198 Madison Avenue, New York, New York 10016

Oxford is a registered trademark of Oxford University Press

Library of Congress Cataloging-in-Publication Data
Liebovitch, Larry S.
    Fractals and chaos simplified for the life sciences / Larry S.
Liebovitch.
        p.   cm.
    Includes bibliographical references and index.
    ISBN 0-19-512024-8 (pbk.)
    1. Medicine—Mathematics.   2. Biology—Mathematics.   3. Fractals.
4. Chaotic behavior in systems.   I. Title.
R853.M3L54   1997
570'.1'51474—dc21   97–20864
                                                           CIP

9 8 7 6 5 4 3 2 1

Printed in the United States of America
on acid-free paper

"Chance favors the prepared mind."

-Louis Pasteur

Preface

"Fractals" and "chaos" have attracted wide attention and excitement in mathematics and the physical sciences.

This book explains the properties of fractals and chaos and shows how they are now being used in biology and medicine.

These ideas are presented in a way that is understandable to people who may not be familiar with advanced mathematical concepts.

This is done by presenting one concept at a time on each set of facing pages. The left-hand page is text and the right-hand page is graphics. The text and graphics each explain the same concept in different ways. Sometimes the text and graphics provide similar information. Sometimes they provide complementary information.

Nonspecialists can use this book to gain a basic understanding of fractals and chaos and their importance in biomedical research.

Teachers in high school or college can use this book as the basis for a unit on fractals and chaos. The right-hand graphics pages can be copied onto transparencies to illustrate the concepts described on the facing left-hand pages.

Biomedical scientists can use this book to understand how concepts from fractals and chaos are being used to study the shape and function of proteins, cell membranes, nerve cells, muscle cells, the lung, the heart, and the brain. They can see how to use these methods to analyze their own data and interpret the results.

Acknowledgments

I thank David Axelrod, Gregory Dewey, Zbigniew Grzywna, Leo Levine, Cynthia Park, Monika Sasksena, Daniela Scheurle, Virginia Standish, Angelo Todorov, and Michal Zochowski for their helpful comments and suggestions.

Abbreviated Contents

Contents

Contents

Fractals and Chaos Simplified for the Life Sciences

Contents

Part III   OTHER METHODS

The Big Picture

Fractals and Chaos Simplified for the Life Sciences

Part I
FRACTALS

Objects or processes whose small pieces
resemble the whole.

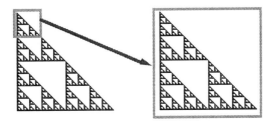

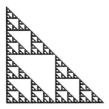

FRACTALS

Introduction

The Difference between Non-Fractal and Fractal Objects

1. Non-Fractal

As a **non-fractal** object is magnified, no new features are revealed.

2. Fractal

As a **fractal** object is magnified, ever finer features are revealed.

The shapes of the smaller features are kind of like the shapes of the larger features.

Non-Fractal

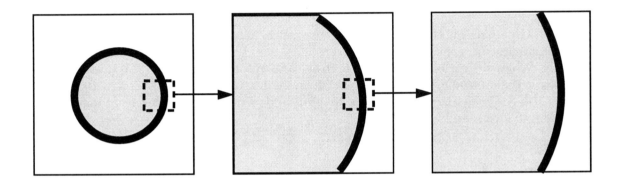

Fractal

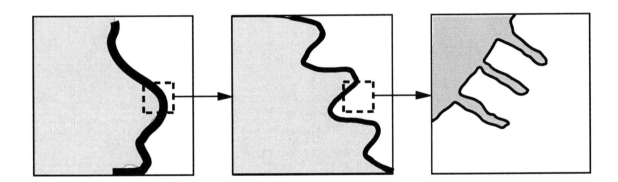

The Sizes of the Features of Non-Fractal and Fractal Objects

1. Non-Fractal

The size of the smallest feature of a **non-fractal** object is called its *characteristic scale.*

When we measure the length, area, and volume at a resolution that is finer than the characteristic scale, then all of the features of the object are included. Thus the measurements at this resolution determine the correct values of the length, area, and volume.

2. Fractal

A **fractal** object has features over a broad range of sizes.

As we measure the length, area, and volume at ever finer resolution, we include ever more of its finer features. Thus the length, area, and volume depend on the resolution used to make the measurement.

6

Non-Fractal

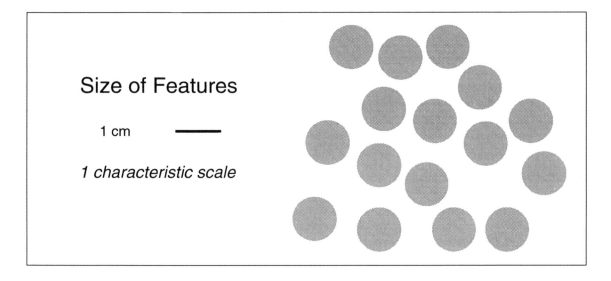

Size of Features

1 cm ————

1 characteristic scale

Fractal

Size of Features

2 cm ————————

1 cm ————

1/2 cm ——

1/4 cm —

many different scales

The Properties of Fractals

1. Self-Similarity

A coastline looks wiggly. You would think that as you enlarge a piece of the coastline the wiggles would be resolved and the coastline would look smooth. But it doesn't. No matter how much you enlarge the coastline it still looks just as wiggly. The coastline is similar to itself at different magnifications. This is called **self-similarity**.

2. Scaling

Because of self-similarity, features at one spatial resolution are related to features at other spatial resolutions. The smaller features are smaller copies of the larger features. The length measured at finer resolution will be longer because it includes these finer features. How the measured properties depend on the resolution used to make the measurement is called the **scaling relationship**.

3. Dimension

The dimension gives a quantitative measure of self-similarity and scaling. It tells us how many new pieces of an object are revealed as it is viewed at higher magnification.

4. Statistical Properties

Most likely, the statistics that you were taught in school was limited to the statistics of non-fractal objects. Fractals have different statistical properties that may surprise you!

Self-Similarity

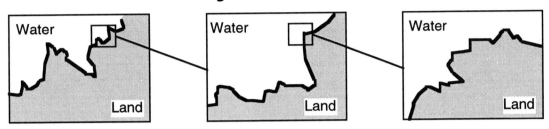

Scaling

The value measured for a property depends on the resolution at which it is measured.

Dimension

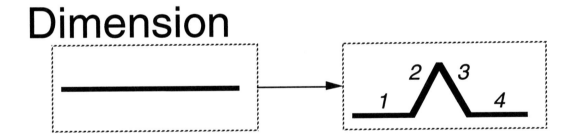

Statistical Properties

moments may be <u>*zero*</u> *or* <u>*non-finite*</u>.

for example,   *mean* $\rightarrow$ *0*
             variance $\rightarrow$ ∞

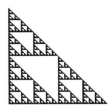

FRACTALS
Self-Similarity

Two Types of Self-Similarity

1. Geometrical Self-Similarity

In **geometrical self-similarity** little pieces of an object are *exact* smaller copies of the whole object.

The little pieces are *geometrically similar* to the whole object.

This is usually true only for mathematically defined objects.

2. Statistical Self-Similarity

The little pieces of real biological specimens are usually not exact smaller copies of the whole object. Thus, they cannot be geometrically self-similar. However, the little pieces of real biological specimens can be kind-of-like their larger pieces. The statistical properties of the little pieces can be geometrically similar to the statistical properties of the big pieces. This is called **statistical self-similarity**.

For example, the statistical property could be the length of the perimeter of an organ. Statistical self-similarity means that the length measured at one resolution is geometrically similar, that is, proportional to the length measured at other resolutions. If $Q(r)$ is the length measured at resolution r, and $Q(ar)$ is the length measured at resolution ar, then $Q(ar) = k\ Q(r)$ where k is the constant of proportionality.

The statistical properties of an object are described by the number of pieces of each size that make up the object. The function that tells how many pieces of each size that make up the object is called the *probability density function (pdf)*. The formal mathematical definition of statistical self-similarity is that the probability density function (pdf) measured at resolution r is geometrically similar, that is, has the same shape, as the probability density function (pdf) measured at resolution ar.

We use the shorter expression "self-similarity" to denote the statistical self-similarity of biological objects in space or processes in time. This means that the smaller pieces are like the larger pieces but they are not exact copies of the larger pieces.

Self-Similarity

Geometrical

The magnified piece of an object is an exact copy of the whole object.

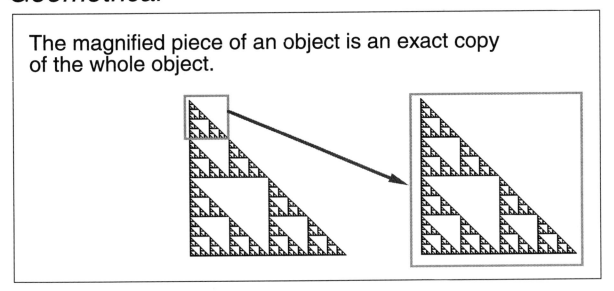

Statistical

The value of statistical property Q(r)
measured at resolution   r,
is proportional to the value Q(ar)
measured at resolution ar.

Q(ar) = k Q(r)

pdf [Q(ar)] $\overset{d}{=}$ pdf [kQ(r)]

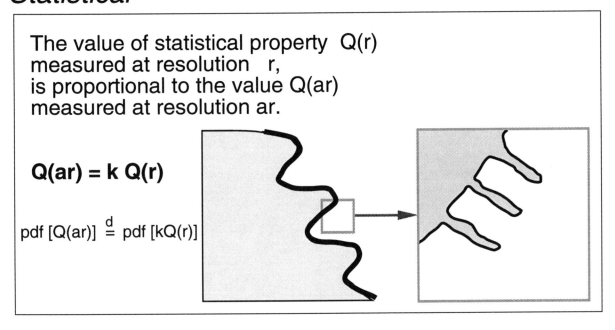

Examples of Self-Similarity in Space

Examples of statistical self-similarity in space are:

1. The branching patterns of the arms (dendrites) of nerve cells growing in the retina in the eye or growing in a nutrient medium in a plastic culture dish,

2. The arteries and veins in the retina, and

3. The tubes that bring air into the lungs.

These objects are self-similar because the pattern of the branching of the large structures is repeated in the branching of the ever smaller structures.

The fractal analysis of these branching patterns sheds light on how the dendrites of the nerve cell grow and how the airways in the lung are formed in the embryo. Determining the fractal properties of the blood vessels in the retina may be useful in diagnosing diseases of the eye or in determining the severity of the disease.

The analysis of nerve cells and blood vessels in the retina and in the nutrient medium of a culture dish is simplified by the fact that the retina and the nutrient media are both very thin. Thus the branching pattern is 2-dimensional. Analysis of a photograph of a 3-dimensional pattern would have to compensate for the images of the branches projected on top of each other.

Branching Patterns

nerve cells
in the retina, and in culture

Caserta, Stanley, Eldred, Daccord, Hausman, and Nittmann 1990
Phys. Rev. Lett. 64:95-98

Smith Jr., Marks, Lange, Sheriff Jr., and Neale 1989
J. Neurosci. Meth. 27:173-180

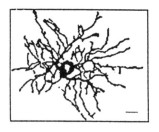 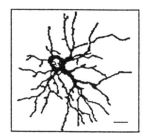 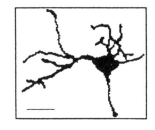

blood vessels
in the retina

Family, Masters, and Platt 1989
*   Physica D38:98-103*
Mainster 1990 Eye 4:235-241

airways
in the lungs

West and Goldberger 1987
Am. Sci. 75:354-365

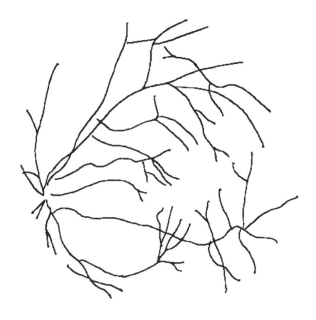 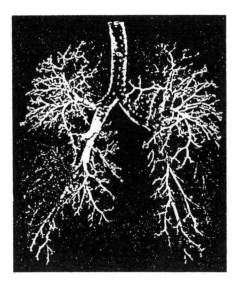

Examples of Self-Similarity in Time

Examples of statistical self-similarity in time are:

1. The average height of a layer of cells above the plastic culture dish in which they are growing, and

2. The electrical voltage across the cell membrane of a T-lymphocyte cell.

These measurements are self-similar because the pattern of the smaller fluctuations over brief times is repeated in the larger fluctuations over longer times.

Variations in Time

cell height above a substrate

Giaever and Keese 1989 Physica D38:128-133

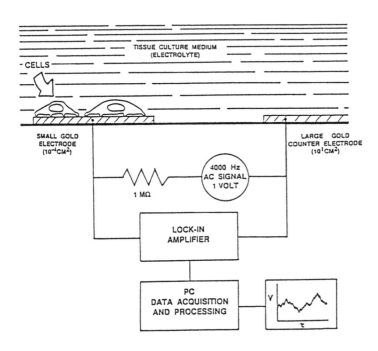

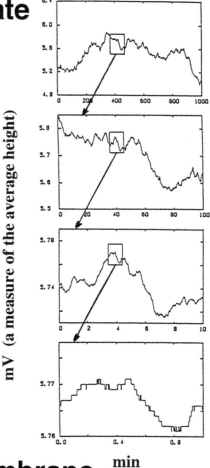

min

voltage across the cell membrane

Churilla, Gottschalke, Liebovitch, Selector, Todorov, and Yeandle 1996
Ann. Biomed. Engr. 24:99-108

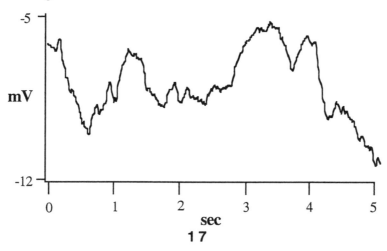

The Currents through Ion Channels
Are Self-Similar in Time

Ion channels are proteins in the cell membrane with a central hole that allows ions such as sodium, potassium, and chloride to get into or out of the cell. The structure of the ion channel protein can change, closing this hole and blocking the flow of ions. The small electrical current (10^{-12} amp) due to these ions can be measured in an individual ion channel molecule. This current is high when the channel is open and low when it is closed.

When a recording of current is played back at low time resolution, the times during which the channel was open and closed can be seen. When one of these open or closed times is played back at higher time resolution, it can be seen to consist of many briefer open and closed times.

The current through the channel is self-similar in time because the pattern of open and closed times found at low time resolution is repeated in the open and closed times found at higher time resolution.

Currents Through Ion Channels

ATP sensitive potassium channel in
β cell from the pancreas

Gilles, Falke, and Misler
(Liebovitch 1990 Ann. N. Y. Acad. Sci. 591:375-391)

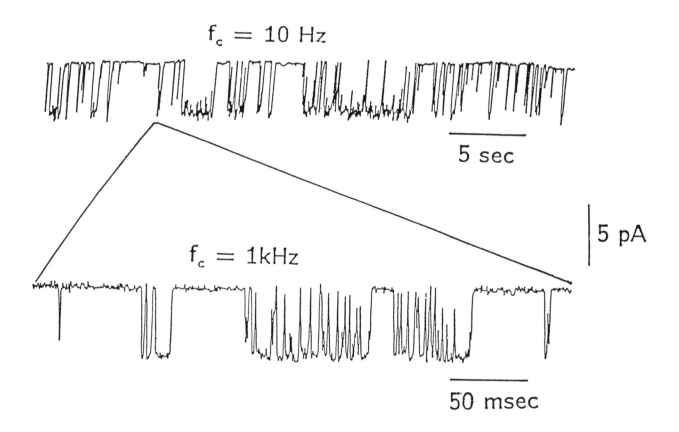

The Open and Closed Times of Ion Channels
Are Statistically Self-Similar

The number of times that the ion channel is closed for each range of closed times is called the *closed time histogram.*

We recorded the current through potassium channels in the cells that line the inside of the cornea in the eye. We played back these current recordings through an analog to digital (A/D) converter into a computer. We then used the computer to determine the histogram of the closed times at these different time resolutions.

Sampling the data at high time resolution by using a high A/D rate best reveals the statistical properties of the brief closed times. Sampling the data at a low time resolution by using a low A/D rate best reveals the statistical properties of the long closed times.

The horizontal time axis is different for histograms computed from the data sampled at different A/D rates. However, these histograms all have the same shape. These statistical distributions of closed times measured at different resolutions in time are geometrically similar. Thus the closed times are statistically self-similar.

Closed Time Histograms

potassium channel in the corneal endothelium

Liebovitch et al. 1987 Math. Biosci. 84:37-68

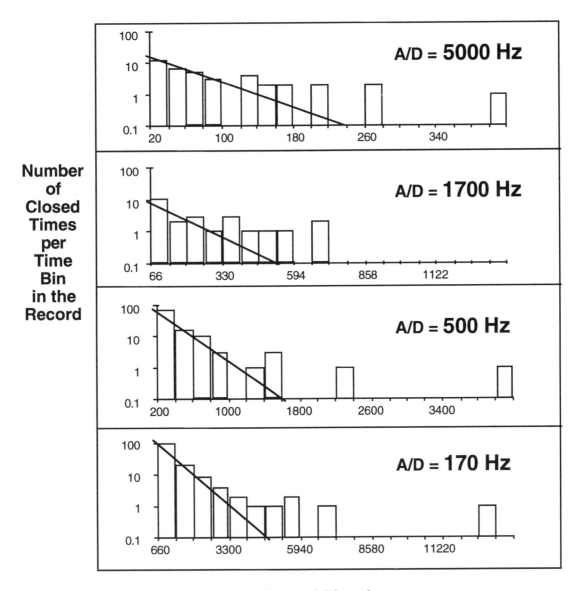

Closed Time in ms

More Examples of Self-Similarity

1. Self-Similarity in Space

We have already seen that there are self-similar patterns in the branching of the arms (dendrites) of nerve cells, the arteries and the veins in the retina, and the tubes that bring air into the lungs.

Additional examples of self-similarity in space are the branching structures of the conduction fibers in the heart (His-Purkinje system) that spread the wave of muscle contraction during the heartbeat; the tubes (ducts) in the liver that bring the bile to the gallbladder; and the arteries and the veins throughout the body.

Many surfaces in the body have self-similar undulations with ever finer fingers or pockets. These ever finer structures increase the area available for the exchange of nutrients, gasses, and ions. These surfaces include the lining of the intestine, the boundary of the placenta, and the membranes of cells.

These properties are self-similar in physical space. There are also self-similar properties in more abstract spaces such as the energy space used to characterize molecules. For example, the distribution of the spacing between energy levels in some proteins is self-similar.

2. Self-Similarity in Time

We have already seen that there are self-similar patterns in time in the motion of the heights of cells above the dish in which they are growing, in the variation of the voltage across the cell membrane, and in the timing of the switches between the open and closed states of an ion protein.

Additional examples of self-similarity in time include the electrical signal generated by the contraction of the heart and the volumes of breaths over time drawn into the lung.

Biological Examples of Self-Similarity

spatial

dendrites of neurons

airways in the lung

ducts in the liver

vessels in the circulatory system

intestine

placenta

cell membrane

energy levels in proteins

temporal

fluctuations in the heights of cell above their substrate

voltage across the cell membrane

timing of the opening and closing of ion channels

heartbeat

volumes of breaths

Biological Implications of Self-Similarity

How can the human genome with only 100,000 genes contain all the information required to construct structures like the heart, which has 1,000,000 capillaries, or the brain, which has 100,000,000,000 nerve cells?

Perhaps the genes do not determine these structures. They may instead determine the rules that generate these structures. Repeated application of these rules may then lead to self-similar structures with many pieces at different resolutions.

How is the body formed?

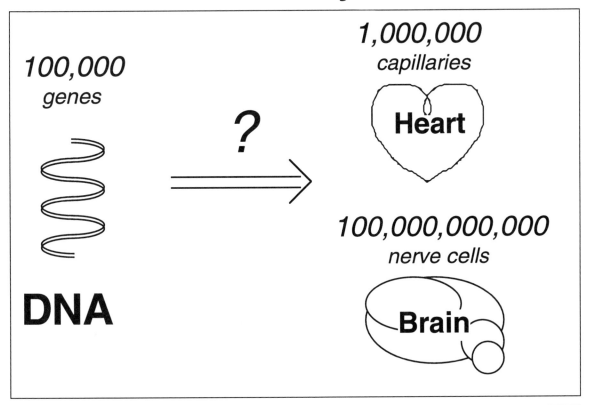

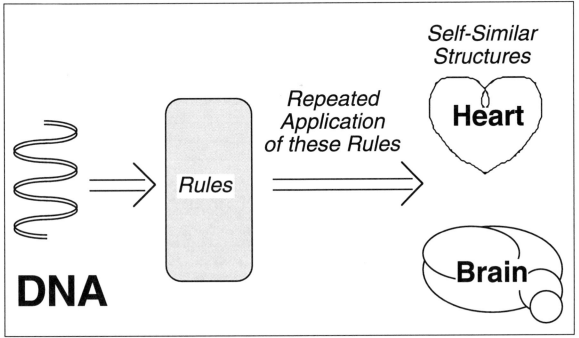

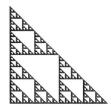

FRACTALS
Scaling

Self-Similarity Implies a Scaling Relationship

Smaller pieces of a fractal will be seen at finer resolution. A measurement made at finer resolution will include more of these smaller pieces. Thus the **value measured** of a property, such as length, surface, or volume, **will depend on the resolution used to make the measurement**. How a measured property depends on the resolution used to make the measurement is called the **scaling relationship**.

Self-similarity specifies how the small pieces are related to the large pieces. Thus self-similarity determines the scaling relationship.

The mathematical form of self-similarity determines the mathematical form of the scaling relationship. The mathematical form of self-similarity is that the value $Q(ar)$ of a property measured at resolution ar is proportional to the value $Q(r)$ measured at resolution r. That is, $Q(ar) = k\ Q(r)$, where k is a constant. From this form, it can be shown that the scaling relationship has one of two possible forms.

1. Power Law

The simplest form of the scaling relationship is that the measured value of a property $Q(r)$ depends on the resolution used to make the measurement with the equation: $Q(r) = B\ r^b$. In this equation B and b are constants. This form is called a *power law*.

2. Full Form

The full form of the scaling relationship is the equation:
$Q(r) = B\ r^b\ f(Log[r]/Log[a])$ where B, b, and a are constants and f(x) is a periodic function such that $f(1+x)=f(x)$.

Self-Similarity $\leftrightarrow$ Scaling

$$Q(ar) = k\, Q(r)$$

$$Q(r) = B\, r^b$$

$$Q(r) = B\, r^b\, f(\text{Log}[r]/\text{Log}[a])$$

Self-Similarity can be satisfied by the power law scaling:

$$Q(r) = B\, r^b$$

Proof:
Using the scaling relationship to evaluate $Q(r)$ and $Q(ar)$,

$$Q(r) = B\, r^b$$

$$Q(ar) = B\, a^b\, r^b$$

If $k = a^b$ then $Q(ar) = k\, Q(r)$

Self-Similarity can be satisfied by the more complex scaling:

$$Q(r) = B\, r^b\, f\left(\frac{\log r}{\log a}\right) \quad \text{where } f(1+x) = f(x)$$

Proof:
Using the scaling relationship to evaluate $Q(a)$ and $Q(ar)$,

$$Q(r) = B\, r^b\, f\left(\frac{\log r}{\log a}\right)$$

$$Q(ar) = B\, a^b\, r^b f\left(\frac{\log ar}{\log a}\right) = B\, a^b\, r^b f\left(\frac{\log a + \log r}{\log a}\right)$$

$$= B\, a^b\, r^b f\left(1 + \frac{\log r}{\log a}\right) = B\, a^b\, r^b f\left(\frac{\log r}{\log a}\right)$$

If $k = a^b$ then $Q(ar) = k\, Q(r)$

Scaling Relationships

The scaling relationship describes how the measured value of a property $Q(r)$ depends on the resolution r used to make the measurement. It can have two different forms.

1. Power Law

The simplest and most common form of the scaling relationship is that $Q(r) = B\ r^b$, where B and b are constants.

On a plot of the *logarithm of the measured property,* **Log [$Q(r)$]**, *versus* the *logarithm of the resolution used to make the measurement,* **Log [r]**, this scaling relationship is a **straight line**.

Such power law scaling relationships are characteristic of fractals.

Power law relationships are found so often because so many things in nature are fractal.

2. Full Form

The full form of the scaling relationship is that
$Q(r) = B\ r^b\ f[1\ +\ Log(r)/Log(a)]$ where B, b, and a are constants and f(x) is a periodic function such that f(1+x)=f(x).

On a plot of the logarithm of the measured property, Log [$Q(r)$], versus the logarithm of the resolution used to make the measurement, Log [r], this scaling relationship is a straight line with a periodic wiggle.

30

Scaling Relationships

most common form:

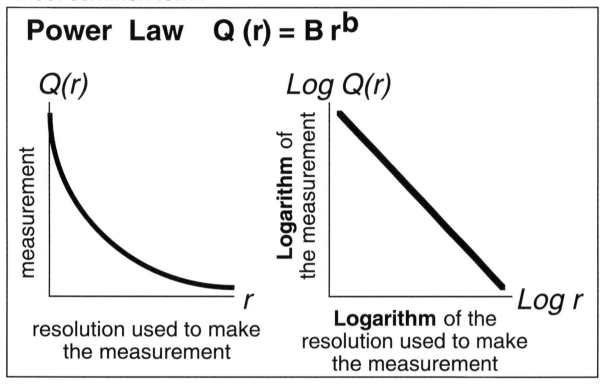

Power Law   Q (r) = B r^b

Q(r)

measurement

resolution used to make
the measurement

r

Log Q(r)

Logarithm of
the measurement

Logarithm of the
resolution used to make
the measurement

Log r

less common, more general form:

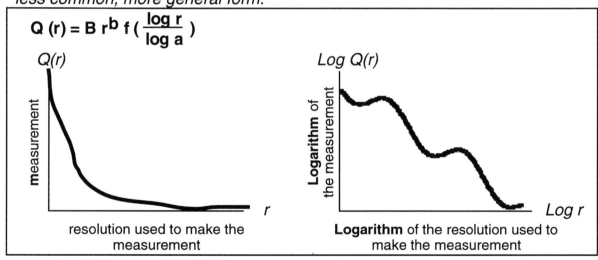

$$Q (r) = B\, r^b\, f\left(\frac{\log r}{\log a}\right)$$

Q(r)

measurement

resolution used to make the
measurement

r

Log Q(r)

Logarithm of
the measurement

Logarithm of the resolution used to
make the measurement

Log r

Example of a Power Law Scaling of a Spatial Object:
The Length of the Coastline of Britain

Richardson measured the length of the coastline of Britain by laying small straight line segments of the same length, end to end, along the coastline. The length of these line segments set the spatial resolution of the measurement. The total length of the coastline was the combined length of all these line segments

When he measured the coastline at finer resolution, the smaller line segments included the smaller bays and peninsulas that were not included in the measurement at coarser resolution. These additional smaller bays and peninsulas increased the total length of the coastline. He found that the length of the coastline was ever longer as he measured it at ever finer resolutions.

He plotted the logarithm of the total length of the coastline, $\text{Log} [Q(r)]$, versus the logarithm of the length of the line segments used to do the measurement, $\text{Log} [r]$. The data were a straight line. He found that $\text{Log } Q(r)$ was proportional to $(-1/4) \text{Log} [r]$. This means that the scaling relationship of the measurement of the length of the coastline has the power law form that $Q(r)$ is proportional to $r^{-(1/4)}$.

How Long is the Coastline of Britain?

Richardson 1961 The problem of contiguity: An Appendix to *Statistics of Deadly Quarrels General Systems Yearbook*   6:139-187.

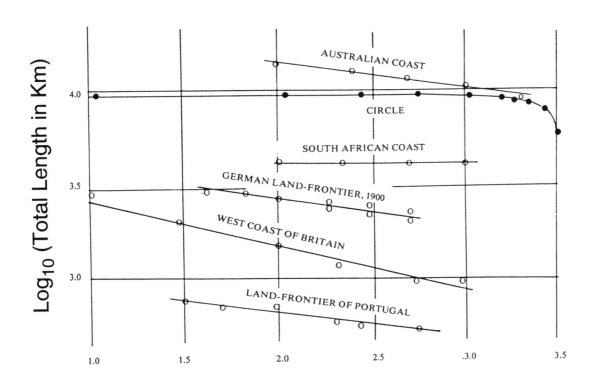

Examples of Power Law Scalings of Spatial Objects: The Surfaces of Cell Membranes

Membranes encase living cells, as well as form different compartments within cells. Paumgartner, Losa, and Weibel measured the total surface area of such membranes from photographs taken with an electron microscope. The magnification of the electron microscope set the spatial resolution of the measurement of the surface area.

As the magnification was increased, then more undulations in the membranes were detected, and the amount of membrane surface measured increased.

They found that the surface area $Q(r)$ measured at resolution r of the endoplasmic reticulum, the outer mitochondrial membrane, and the inner mitochondrial membrane each had a power law scaling of the form that $Q(r)$ was proportional to r^b, where the value of b was different for each type of membrane.

This power law scaling of the membrane surface area $Q(r)$ with the resolution r used to measure it appears as a straight line on a plot of Log $[Q(r)]$ versus Log $[r]$. The slope of these lines, which is equal to the value of b, is different for each type of membrane.

Scaling of Membrane Area

Paumgartner, Losa, and Weibel 1981
J. Microscopy   121:51-63

imi	inner mitochondrial membrane
omi	outer mitochondrial membrane
er	endoplasmic reticulum

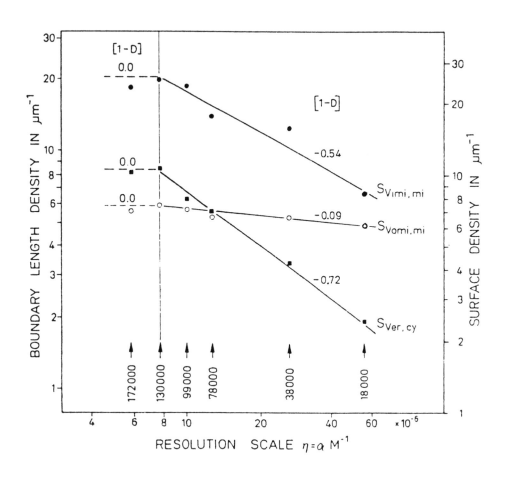

Example of a Power Law Scaling of a Process in Time: Ion Channel Kinetics

We wanted to find out if there is a scaling relationship present in the open and closed times that we recorded from a potassium ion channel in the cells that line the inside of the cornea in the eye.

We needed to determine how the probability to change from closed to open depends on the time resolution used to measure it. The channel must be closed long enough for us to see it as closed. Thus, from our data, we determined the *conditional probability* that a closed channel will open, *given* that it has already remained closed *for a certain amount of time.* That certain amount of time then sets the time resolution at which the measurement is done.

We called this conditional probability the *effective kinetic rate constant,* k_{eff} and the time resolution at which it was measured the *effective time scale,* t_{eff}. We plotted the logarithm of the effective kinetic rate constant Log [k_{eff}] versus the logarithm of the effective time scale Log [t_{eff}] at which it was measured. The data were a straight line on this plot. That tells us that the ion channel kinetics has a power law scaling characteristic of a fractal, where k_{eff} is proportional to t_{eff}^{b}.

Scaling of Ion Channel Kinetics

Liebovitch et al. 1987 Math. Biosci. 84:37-68

70 pS Channel, on cell, Corneal Endothelium

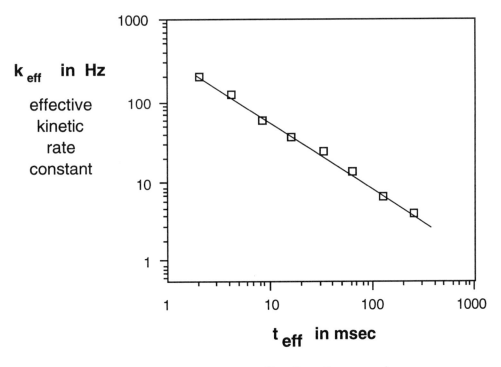

effective time scale

The Physical Significance of the Scaling
of Ion Channel Kinetics

The scaling relationship in the timing of the switching between the open and closed states could be interpreted in two different ways:

1. It could reflect the energy **structure** of the ion channel protein. The channel protein could exist in a large number of similar, but not identical, shapes. In this interpretation, the scaling relationship can be used to determine the heights of the energy barriers that separate the closed shapes from their corresponding open shapes.

2. It could reflect the **dynamics** of how the shape of the ion channel protein changes in time. In this interpretation, the scaling relationship can be used to determine how the energy barrier between the closed state and the open state varies in time.

The fractal approach focused attention on the dynamics, that is, the motions of the structures within ion channel proteins, on how past motions affect future behavior, and on the importance of the large number of slightly different shapes of the channel protein.

The results of 20 years of experiments on other proteins also suggested that internal motions and the large numbers of possible protein shapes play an important role in how all proteins work.

In contrast, most scientists who studied only ion channels had analyzed their data as if ion channel proteins had only a few states with fixed structures. Our work made some of these scientists very unhappy. Scientists, like anyone else, sometimes react with hostility to new ideas that challenge their beliefs and their previous work.

Two Interpretations
of the Fractal Scalings

Structural

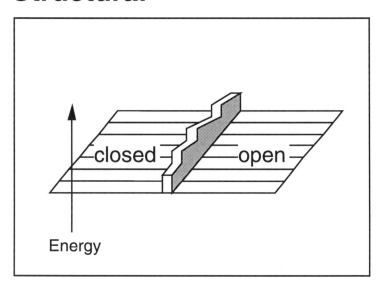

The scaling relationship reflects the distribution of the activation energy barriers between the open and closed sets of conformational substates.

Dynamical

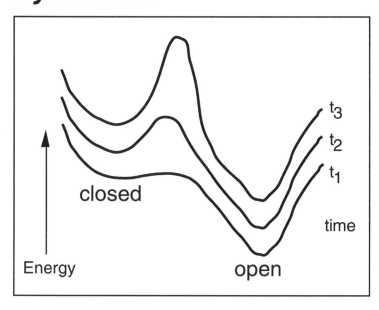

The scaling relationship reflects the time dependence of the activation energy barrier between the open and closed states.

More Examples of Scaling Relationships

1. Scaling Relationships in Space

We have already seen that there is a scaling relationship that tells us how the amount of membrane area measured in a cell depends on the resolution in space used to measure it.

Additional examples of scaling relationships in space include how the diameter of the tubes that bring air into the lung, the width of the spaces between the cells that line the capillaries in the lung, and the surface areas of proteins each depend on the spatial resolution used to measure them.

2. Scaling Relationships in Time

We have already seen that there is a scaling relationship that tells us how the rate of switching between the open and closed states of an ion channel protein depends on the resolution in time used to measure it.

Another example of scaling relationships in time is the rate of chemical reactions that are limited by the time it takes for the molecules to reach each other. The reaction slows down with time because the nearby molecules have reacted and it takes ever longer for the unreacted molecules to reach each other. The scaling relationship describes how the kinetic rate constant of the reaction depends on the time measured since the reaction began.

A scaling relationship in time is also present in the washout kinetics of how the concentration of substances in the blood decay with the time measured since the substances were first injected into the blood.

Biological Examples of Scaling Relationships

spatial

area of endoplasmic reticulum membrane

area of inner mitochondrial membrane

area of outer mitochondrial membrane

diameter of airways in the lung

size of spaces between endothelial cells in the lung

surface area of proteins

temporal

kinetics of ion channels

reaction rates of chemical reactions limited by diffusion

washout kinetics of substances in the blood

Biological Implications of Scaling Relationships

1. No Unique, "Correct" Value for a Measurement

The scaling relationship tells us how the value measured for a property, such as length, area, or volume, depends on the resolution used to make the measurement. There is no one value that represents the "correct" value of the property being measured. **The value measured for a property depends on the resolution used to make the measurement.**

2. Measurements Made at Different Resolutions Will Be Different

The value measured for a property depends on the resolution used to make the measurement. Thus measurements made at different resolutions will yield different values. This means that the **differences between the values measured by different people could be due to the fact that each person measured the property at a different resolution.** It also means that it is very important to state the resolution at which a measurement is performed.

3. Importance of the Scaling Relationship

Our method of analyzing the data must be consistent with the characteristics of the data if the results are to be meaningful. Only if the analysis properly characterizes the data can it give us clues about the nature of the process that produced the data. **The measurement of the value of a property at only one resolution is not useful** to characterize fractal objects or processes. Instead, we need to determine **how the values measured for a property depend on the resolution used to make the measurement,** namely, the **scaling relationship.** This change, from measuring a single value to measuring how the values depend on the resolution is called a change in *paradigm.*

Scaling

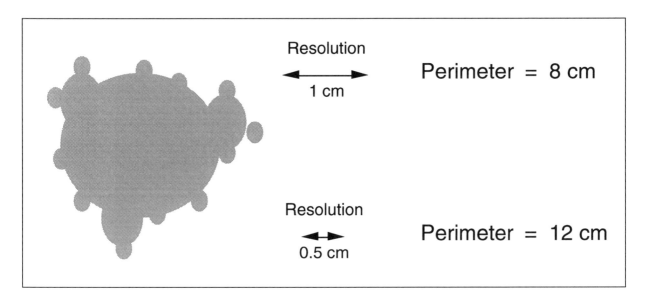

Resolution
1 cm

Perimeter = 8 cm

Resolution
0.5 cm

Perimeter = 12 cm

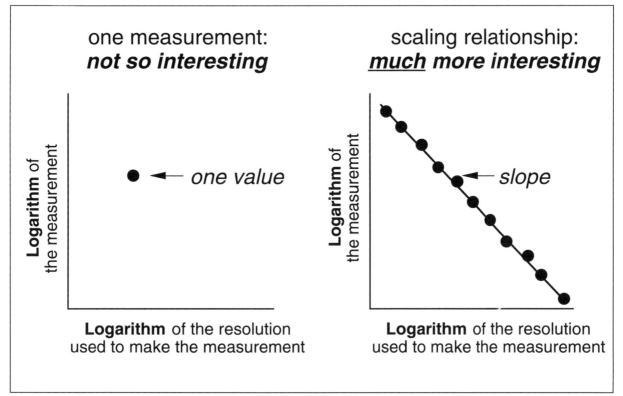

one measurement:
not so interesting

scaling relationship:
much more interesting

Logarithm of
the measurement

← *one value*

Logarithm of
the measurement

← *slope*

Logarithm of the resolution
used to make the measurement

Logarithm of the resolution
used to make the measurement

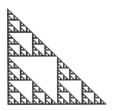

FRACTALS
Dimension

Dimension: A Quantitative Measure of Self-Similarity and Scaling

The **dimension** gives a quantitative measure of the fractal properties of self-similarity and scaling.

The dimension tells us how many additional smaller pieces of an object are revealed when it is magnified by a certain amount.

There are a number of different dimensions. Each dimension measures different properties of an object. They can be grouped into three different types of dimensions:

1. Fractal Dimension

The fractal dimension describes how an object fills up space.
It gives us information about the length, area, or volume of an object.
Its value can be an *integer* or a *fraction* (for example: 2 or 1.7).

2. Topological Dimension

The topological dimension describes how points within an object are connected to each other.
It tells us that an object is an edge, a surface, or a solid.
Its value is always an *integer* (for example: 1 or 2).

3. Embedding Dimension

The embedding dimension describes the space that contains the object.
It tells us that this space is a line, an area, or a volume.
Its value can be an *integer* or a *fraction* (for example: 2 or 1.7).

Dimension

A quantitative measure of self-similarity and scaling.

> **The dimension tells us how many new pieces we see when we look at finer resolution.**

Fractal Dimension

space filling properties of an object

e.g.   self-similarity dimension
       capacity dimension
       Hausdorff-Besicovitch dimension

Topological Dimension

how points within an object are connected

e.g.   covering dimension
       iterative dimension

Embedding Dimension

the space that contains an object

The Simplest Fractal Dimension:
The Self-Similarity Dimension

The **fractal dimension describes the space filling properties of an object**. There are many different fractal dimensions. Each one characterizes the space filling properties of an object in a slightly different way. The simplest fractal dimension is called the **self-similarity dimension**.

Consider a geometrically self-similar fractal object made up of line segments. To evaluate the self-similarity dimension we *divide each line segment into M smaller line segments*. This will produce *N smaller objects*. If the object is geometrically self-similar, each of these smaller objects is an exact but reduced size copy of the whole object. The self-similarity dimension **d** is then found from the equation $N = M^d$. This equation can also be written as **d = Log (N) / Log (M)**.

For example, when M = 3, we replace each line segment of an object with 3 little line segments.

If we divide a line into thirds, it produces 3 little lines. Thus N = 3. The equation $N = M^d$ has the form $3 = 3^1$. Hence, the self-similarity dimension d of the line is equal to **1**.

If we divide each side of a square into thirds, it produces 9 little squares. Thus N = 9. The equation $N = M^d$ has the form $9 = 3^2$. Hence, the self-similarity dimension d of the square is equal to **2**.

If we divide each side of a cube into thirds, it produces 27 little cubes. Thus N = 27. The equation $N = M^d$ has the form $27 = 3^3$. Hence, the self-similarity dimension d of the cube is equal to **3**.

These dimensions computed from the self-similarity dimension are consistent with our intuitive idea that the dimensions of length, area, and volume should be 1, 2, and 3.

Fractal Dimensions
Self-Similarity Dimension

N new pieces when
each line segment
is divided by M.

$$N = M^d$$

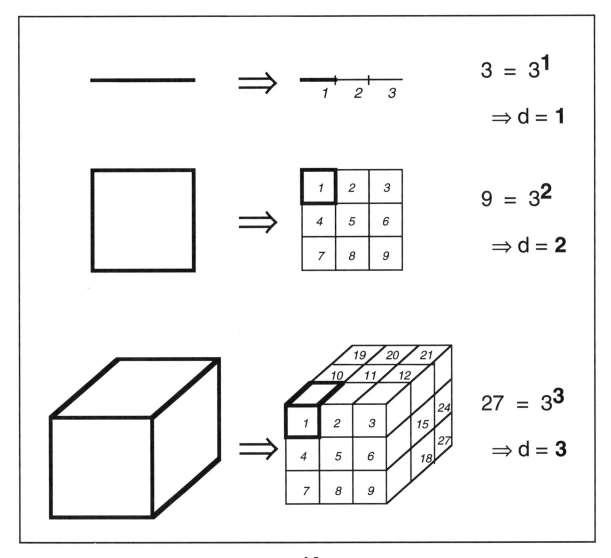

$3 = 3^1$

$\Rightarrow d = 1$

$9 = 3^2$

$\Rightarrow d = 2$

$27 = 3^3$

$\Rightarrow d = 3$

More General Fractal Dimensions:
The Capacity Dimension

The self-similarity dimension requires that each little object formed by dividing the line segments of the whole object into smaller pieces must be an exact copy of the whole object. Thus the self-similarity dimension can only be used to analyze objects that are geometrically self-similar. To determine the dimension of irregularly shaped objects requires a more general form of the fractal dimension. Two such forms are the **capacity** and the **Hausdorff-Besicovitch** dimension.

To evaluate the **capacity** of an object we cover it with "balls" of a certain radius r. We find the smallest number of balls N(r) needed to cover all the parts of an object. We then shrink the radius of the balls and again count the smallest number needed to cover the object. The capacity is the value of **Log N(r) / Log (1/r)** in the limit as the radius of the balls shrinks to 0.

The capacity is a generalization of the self-similarity dimension. The self-similarity dimension d = Log N / Log M, where N is the number of smaller copies of the whole object seen when each line segment is replaced by M pieces. The spatial resolution in the self-similarity dimension is proportional to 1/M. The spatial resolution in the capacity is proportional to the radius r of the balls used to cover the object. Thus r is proportional to 1/M. To arrive at the capacity from the self-similarity dimension, we first replace M by 1/r. This leads us to d = Log N / Log (1/r). Second, instead of counting N, the number of smaller copies of the whole object, we count N(r) the number of balls needed to cover the object. This leads us to d = Log N(r) / Log(1/r). Last, we take the limit of Log N(r) / Log(1/r) as the radius of the balls shrinks to 0.

Fractal Dimensions
Capacity Dimension

N(r) balls of radius r
needed to cover the object.

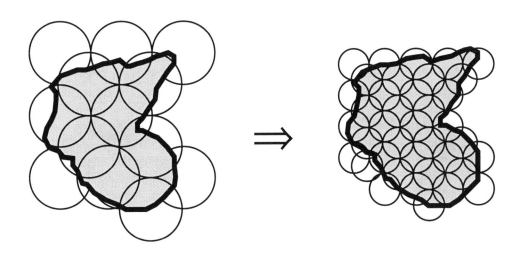

$$d = \lim_{r \to 0} \frac{\text{Log } N(r)}{\text{Log }\left(\frac{1}{r}\right)}$$

Relationship to self-similarity dimension: M=1/r, then N=M^d

More General Fractal Dimensions:
The Hausdorff-Besicovitch Dimension

The Hausdorff-Besicovitch dimension is usually what mathematicians mean when they say "fractal dimension."

The formal definition of the Hausdorff-Besicovitch dimension is quite technical. Only a brief hint of the those details will be given here.

The Hausdorff-Besicovitch dimension is similar, but not identical to, the capacity. In the capacity, we count the number of balls $N(r)$ of a given radius r needed to cover an object. The equation $d = \text{Log } N(r) / \text{Log } (1/r)$, implies that $N(r)$ the number of balls needed to cover the object is proportional to r^{-d}. The capacity dimension d is determined directly from how the number of balls needed to cover the object varies with the radius of the balls.

In the **Hausdorff-Besicovitch dimension**, we cover the object with sets. We evaluate the sum of a function applied to the diameter of each covering set. This function is called the gauge function. Analogous to the capacity and self-similarity dimension, the gauge function that is used is to raise the diameter r of each set to the power s. The sum of the diameters of all the sets each raised to the power s is then computed. The behavior of this sum as a function of s is then studied in the limit as the diameter r of the sets approaches 0. As r approaches 0, this sum will grow very large if s is less than a certain number, and it will grow very small if s is greater than a certain number. The value of the number that separates these two types of behavior is called the Hausdorff-Besicovitch dimension.

We can now see how the Hausdorff-Besicovitch dimension is similar to the capacity. The number of sets needed to cover an object is proportional to r^{-d}, where d is the capacity. The number of sets times the diameter of each raised to the power s is thus equal to r^{s-d}. As r approaches 0, this sum will grow very large if $s<d$, and it will grow very small if $s>d$. Thus the boundary between these limits occurs when $s = d$.

Fractal Dimensions
Hausdorff-Besicovitch Dimension

A_i = covering sets

$$H(s,r) = {}^{\inf} \sum_i (\text{diameter } A_i)^s$$

$$\lim_{r \to 0} H(s,r) = \infty \quad \text{for all } s < d$$

$$\lim_{r \to 0} H(s,r) = 0 \quad \text{for all } s > d$$

Example of Determining the Fractal Dimension:
Using the Self-Similarity Dimension

The Koch curve is constructed by starting with an equilateral triangle. At each stage in the construction, each line segment is divided into thirds. The two end pieces are left intact. Each middle piece is then replaced by two pieces. Each of the 4 pieces is 1/3 as long as the original line segment. This procedure can be repeated forever.

Each original side of the triangle is 3 units long. After the first stage of construction each side is 4 units long. At each stage in the construction, the length of the perimeter of the Koch curve increases by 4/3. When there is an infinite number of stages in the construction, then the perimeter of the Koch curve is infinitely long.

The perimeter of the Koch curve is geometrically self-similar. At each stage in the construction the original segment was a straight line, and the new smaller segments are also straight lines. Thus the smaller pieces are geometrically similar to the original segment.

We can use the **self-similarity dimension** to determine the dimension of the perimeter of the Koch curve.

As the spatial resolution is increased by a factor of 3, we see 4 new pieces. That is, when the size of the lines making up the perimeter is reduced by 1/3 of their original length, then we find 4 small pieces. The self-similarity dimension d is the logarithm of the number of new pieces divided by the logarithm of the factor of the reduction in size of each piece. Thus $d = \text{Log}(4) / \text{Log}(3) = 1.2619$.

Fractal Dimension of the Perimeter of the Koch Curve

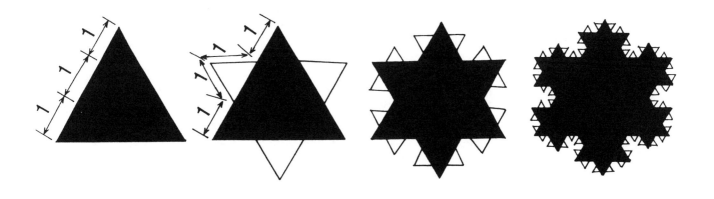

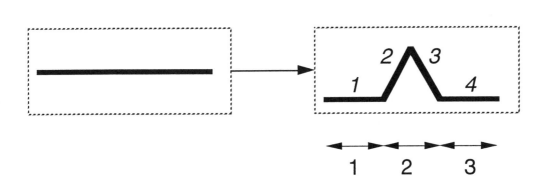

When we look at   3x   finer resolution, we see   4   additional smaller pieces.

Self-Similarity Dimension:

$$d = \frac{\text{Log (number of new pieces)}}{\text{Log (factor of finer resolution)}} = \frac{\text{Log } 4}{\text{Log } 3} = 1.2619$$

Example of Determining the Fractal Dimension:
Using the Capacity Dimension and Box Counting

The capacity dimension d = Log N(r) / Log (1/r), in the limit where r approaches 0, where N(r) is the smallest number of balls of radius r needed to cover an object.

A useful way to evaluate the capacity is to use "balls" that are the boxes of a rectangular coordinate grid. This method is called **box counting**.

For example, we cover an object with a grid and count how many boxes of the grid contain at least some part of the object. We then repeat this measurement a number of times, each time using boxes with sides that are 1/2 the size of the previous boxes.

The capacity dimension is then the slope of the plot of Log N(r) versus Log (1/r), or equivalently, the negative of the slope of the plot of Log N(r) versus Log (r).

If an object is self-similar, then the slope of Log N(r) versus Log (1/r) is the same as the limit of Log N(r) / Log (1/r) as r approaches 0. It is much easier to determine the slope than the limit.

New algorithms make it possible to determine efficiently the number of boxes that contain at least some part of the object. Using these new algorithms, box counting is a particularly good method to evaluate the fractal dimension of images in photographs.

Fractal Dimension of an Object by Box Counting

r = Box Size

$N(r)$ = Number of Boxes Needed to Cover the Set

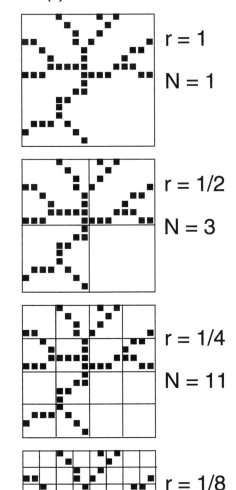

r = 1

N = 1

r = 1/2

N = 3

r = 1/4

N = 11

r = 1/8

N = 26

$N(r) = 1.03\, r^{-1.60}$

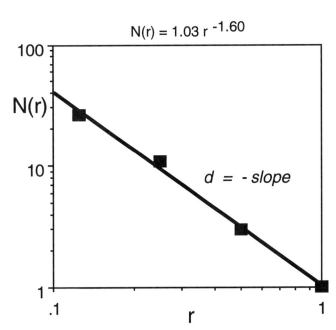

d = - slope

$$d = \frac{\Delta \log N(r)}{\Delta \log (1/r)}$$

$$= -\frac{\Delta \log N(r)}{\Delta \log (r)} = 1.60$$

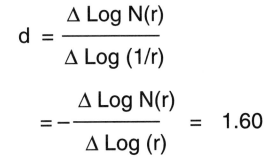

Fast Box Counting Algorithms:

Liebovitch & Tóth 1989 Phys. Lett. A141:386-390.

Hou et al. 1990 Phys. Lett. A151:43-46.

Block et al. 1990 Phys. Rev. A42:1869-1874.

Example of Determining the Fractal Dimension:
Using the Scaling Relationship

The **fractal dimension** can be found from the **scaling relationship**. This is the most often used method to determine the fractal dimension from experimental data.

The fractal dimension d means that the number of pieces $N(r)$ measured at resolution r is proportional to r^{-d}. The scaling relationship evaluated from the experimental data tells us that the value of a property $Q(r)$ measured at resolution r is proportional to r^b. If we know how the property $Q(r)$ depends on the number of pieces and the size of each piece, we can determine the fractal dimension d from the exponent b in the scaling relationship.

For example, Richardson measured the length of the coastline of Britain by laying line segments of length r end to end along the coastline. The length r of the line segments set the resolution used to make the measurement. The total length of the coastline $Q(r)$ is equal to the number $N(r)$ of these line segments multiplied by the length r of each one. That is, $Q(r) = rN(r)$. The definition of the fractal dimension is that the number of line segments $N(r)$ is proportional to r^{-d}. Thus the total length of the coastline $Q(r)$ is proportional to r^{1-d}. Richardson repeated the measurement using line segments of different size r. In this way he determined the scaling relationship that the total length of the coastline $Q(r)$ was proportional to $r^{-.25}$. Equating the exponent of the scaling relationship to the exponent determined from the properties of the dimension, we find $-.25 = 1 - d$. Thus the fractal dimension d of the length of the coastline is equal to 1.25.

Fractal Dimension Determined from the Scaling Relationship

in general:

$$N(r) \propto r^{-d}$$

dimension: $N(r)$ is the number of pieces found at resolution r

$$Q(r) \propto r^{b}$$

scaling relationship of the property $Q(r)$ measured from the data

$$Q(r) \propto [N(r)]^{\alpha} [r]^{\beta}$$

theory on how property $Q(r)$ depends on $N(r)$ and r

Thus, the dimension:

$$d = \frac{\beta - b}{\alpha}$$

for the length of the coastline of Britain:

$$N(r) \propto r^{-d}$$

$$Q(r) \propto r^{-.25}$$

Richardson's measurement of the length of the coastline

$$Q(r) \propto N(r)\, r \propto r^{1-d}$$

The length is the number of line segments times the length of each line segment.

Thus, the dimension:

$$d = 1.25$$

The Topological Dimension

The **topological dimension describes how the points that make up an object are connected together**. The value of the topological dimension is always an *integer*. Edges, surfaces, and volumes have topological dimension of 1, 2, and 3.

For example, we have already seen that the space filling properties of the perimeter of the Koch curve are described by the fractal dimension of about 1.2619. But no matter how wiggly this perimeter is, it is still a line. Thus the topological dimension of the perimeter of the Koch curve is equal to 1.

There are a number of different ways to determine the topological dimension.

1. Covering Dimension

To evaluate the **covering dimension** we first find the least number of sets needed to cover all the parts of an object. These sets may need to overlap each other. If each point of the object is covered by no more than G sets, then the covering dimension $d = G - 1$.

For example, if the least number of circles are used to cover a plane, each point in the plane will be covered by no more than 3 circles. Since $3 - 1 = 2$, then the covering dimension of the plane is 2.

2. Iterative Dimension

The **iterative dimension** is based on the fact that a space of dimension D has borders that have dimension $D-1$. For example, a 3-dimensional volume can be surrounded by 2-dimensional planes. To evaluate the iterative dimension, we find how many times we need to take the borders of the borders of the borders . . . of an object to reach a zero-dimensional point. If we need to repeat the border taking H times, then the iterative dimension is $d = H$.

For example, a plane can be surrounded by borders that are lines. The endpoints of the lines can be delimited by points. We need to take the borders of the borders twice to reach the points. Thus the iterative dimension of the plane is 2.

60

Topological Dimensions

always an *integer*

Covering Dimension

In a minimal covering, each point of the object is covered by no more than G sets.

$$d = G - 1$$

for a plane: $G = 3$ $d = 3 - 1 = 2$

Iterative Dimension

Borders of D dimensional space have dimension D - 1.
Find the borders of the borders.
Repeat H times until 0 dimensional.

$$d = H$$

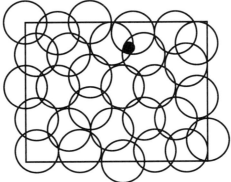

for a plane: *step #1*   *step #2*

$d = 2$

The Embedding Dimension

The embedding dimension describes the fractal dimension of the space that contains the fractal object.

Usually, fractal objects are present in spaces whose dimensions are 1, 2, or 3. For example, the points in time when an ion channel switches open or closed are embedded along a 1-dimensional time line. The retina is very thin. Thus, the nerves and the blood vessels in the retina are embedded in a 2-dimensional space. The tubes that bring air into the lung are spread out into a 3-dimensional space.

However, fractal objects can also be present inside other fractal objects. For example, there are fractal patterns of the motion of molecules along the fractal surface of a solid catalyst.

Embedding Dimension

Fractals can live inside integer dimension spaces:

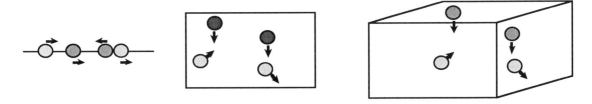

A chemical reaction can occur in 1, 2, or 3 dimensional space.

Fractals can live inside non-integer dimension spaces:

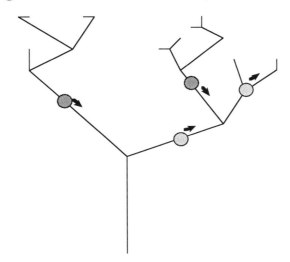

A chemical reaction can also occur in a fractal dimensional space.

Definition of a Fractal

> **A fractal is an object in space or a process in time that has a fractal dimension greater than its topological dimension.**

For example, the perimeter of the Koch curve has a *fractal dimension* of about 1.2619. The fractal dimension describes the space filling properties of the perimeter. The fractal dimension of the perimeter is between 1 and 2. Since the fractal dimension is larger than 1, the perimeter is so wiggly that it covers more than just a 1-dimensional line. But since the fractal dimension is less than 2, the perimeter is not so wiggly that it covers a 2-dimensional area.

The *topological dimension* of the perimeter is 1. The topological dimension describes how the points on the perimeter are connected together. No matter how wiggly the perimeter is, it is still a line with a topological dimension equal to 1.

Since the fractal dimension of the perimeter (1.2619) is greater than the topological dimension of the perimeter (1), the perimeter of the Koch curve is a fractal.

The topological dimension tells us what kind of thing an object is, such as an edge, a surface, or a volume. When the fractal dimension is larger than the topological dimension, it means that the edge, surface, or volume has more finer pieces than we would have expected of an object with its topological dimension. It is more wiggly than we expected. That is why we keep seeing more smaller pieces when we examine it at finer resolution.

The additional smaller pieces at finer resolution mean that the object covers more space than we would have expected of an object with its topological dimension. The topological dimension is an integer. The additional space covered means that the fractal dimension is an integer plus an additional fraction.

Mandelbrot says that he coined the word **"fractal"** to reflect these central ideas that a fractal is: (1) *fragmented* into ever finer pieces and (2) has a *fractional* dimension.

Definition of a Fractal

---

d (fractal) > d (topological)

---

example:

perimeter:

d (fractal)   =   1.2619...

d (topological) =   1.

1.2619... > 1.

d (fractal) > d (topological)

perimeter: covers more space than a 1-D line

covers less space than a 2-D area

"FRACTAL"   fragmented, many pieces

fractional dimension

Example of the Fractal Dimension:
Blood Vessels in the Lungs

Many parts of the body have fractal structures. The fractal dimension can be used to measure the differences between normal structures and those altered by disease.

Laboratory rats that are brought up breathing less oxygen or more oxygen than is found in normal air have higher blood pressure in the blood vessels in the lungs. X-ray images can be taken of these blood vessels. We digitized these images and used the box counting method to determine the fractal dimension of these blood vessels.

We found that the fractal dimension of the blood vessels in the normal lungs was 1.65. We also found that the fractal dimension of the blood vessels in the abnormal lungs was 1.53 and 1.43. That is, the fractal dimension was lower in the abnormal lungs. The fractal dimension tells us how many additional branches are found as smaller blood vessels are examined. Thus there were fewer, finer branches in the blood vessels from the abnormal lungs. The fractal dimension is one way to measure the differences between these normal and abnormal lungs.

Pulmonary Hypertension
HIGH BLOOD PRESSURE IN THE LUNGS

Boxt, Katz, Czegledy, Liebovitch, Jones, Esser & Reid
1994 J. Thoracic Imaging 9:8-13

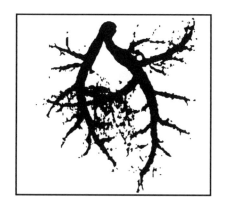

D = 1.65

normal

20% O_2

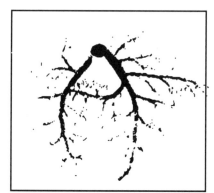

D = 1.53

hypoxic

10% O_2

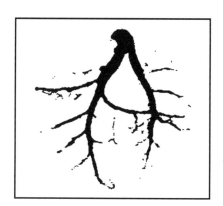

D = 1.43

hyperoxic

90% O_2

More Examples of the Fractal Dimension

We have already seen that the fractal dimension can be used to measure the branching pattern of the blood vessels in the lung. Fractal dimensions have also been measured for many structures and processes at the molecular, cellular, and organ levels.

The most commonly used methods for determining these fractal dimensions have been the use of the scaling relationship and the capacity dimension done by box counting.

Examples where the fractal dimension has been measured include:

1. The *surfaces* of proteins, cell membranes, cells of the cornea damaged by infection with the herpes simplex virus, and bacterial colonies;

2. The *branching patterns* of one type of lipid within the rest of the lipids in the cell membrane, the arms (dendrites) of nerve cells, and the blood vessels in the eye, heart, and lung;

3. The *distribution in space* of the blood flow in the heart, the density of bone and teeth, and the concentration of radioactive tracer in the liver;

4. The *distribution in time* of the intervals between the electrical impulses in nerve cells and the opening and closing of ion channels;

5. The *distribution in energy space* of the differences in energies of vibration in proteins; and

6. The *concentration dependence* of the rates of chemical reactions of protein enzymes.

Biological Examples where the Fractal Dimension has been Measured

surfaces of proteins

surface of cell membranes

shape of herpes simplex ulcers in the cornea

growth of bacterial colonies

islands of types of lipids in cell membranes

dendrites of neurons

blood vessels in the eye, heart, and lung

blood flow in the heart

textures of X-rays of bone and teeth

texture of radioisotope tracer in the liver

action potentials from nerve fibers

opening and closing of ion channels

vibrations in proteins

concentration dependence of reaction rates of enzymes

Biological Implications of the Fractal Dimension

1. Quantitative Measure of Self-Similarity

The numerical value of the fractal dimension d gives us a quantitative measure of self-similarity. The fractal dimension tells us how many small pieces $N(r)$ are revealed when an object is viewed at finer resolution r. The quantitative relationship is that $N(r)$ is proportional to r^{-d}. The larger the fractal dimension, the larger the number of small pieces are revealed as the object is viewed at finer resolution.

2. Quantitative Measure of Correlations

The small pieces of a fractal are each related to the large pieces. Thus the small pieces are also related to each other. The fractal dimension measures the correlations between the small and large pieces. Thus it also measures the correlations between the small pieces themselves.

3. Quantitative Classification

The numerical value of the fractal dimension can be used to classify different objects. For example, the fractal dimension of the blood vessels in a normal retina is different from the fractal dimension in a retina changed by disease. The fractal dimension may serve as a method to diagnose different diseases and as an index to quantify the severity of these diseases.

4. Hints About Mechanisms

Different mechanisms produce fractals with different dimensions. Hence, measuring the fractal dimension of an object may give us clues about figuring out the mechanism that produced it. For example, the process called diffusion limited aggregation produces fractals with a fractal dimension of about 1.7. The blood vessels in the retina have a fractal dimension of about 1.7. Thus it is worthwhile to consider if the growth of these blood vessels was produced by diffusion limited aggregation. This would mean that the growth of the blood vessels was proportional to the gradient of a diffusible substance, such as oxygen or a growth factor.

Fractal Dimension

Numerical Measure of Self-Similarity

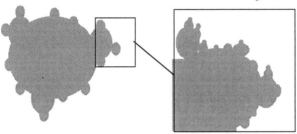

Numerical Measure of Correlations in Space or Time

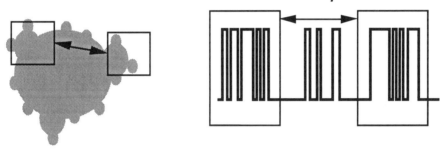

Numerical Measure of Normal versus Sick

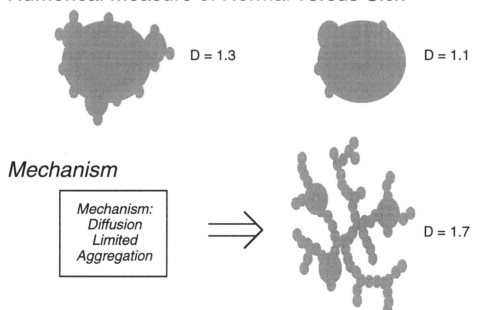

D = 1.3   D = 1.1

Mechanism

Mechanism:
Diffusion
Limited
Aggregation

⟹

D = 1.7

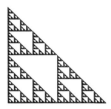

FRACTALS
Statistical Properties

The Statistical Properties of Fractals

Fractals may seem to have strange statistical properties. These properties really aren't strange. They only appear to be strange because they are not covered in the usual statistics courses. Those courses only cover the properties of statistical distributions that are called **Gaussian** or **asymptotically Gaussian** distributions. The statistical properties of fractals belong to a more general class of distributions which are called **stable** distributions. Although few scientists seem to be aware of the properties of stable distributions, mathematicians have studied them for over 250 years.

For example, **the average of a fractal may not exist**. How can the average not exist? If we have 5 measurements, can't we just punch those 5 values into a hand calculator, determine their sum, divide the sum by 5, and get the average?

That calculation on the hand calculator alone is not enough to demonstrate that the average exists. The average from one set of data that we determine with our hand calculator is called the *sample mean*. To show the average exists we must show that as more data are analyzed, these sample means reach a limiting value. We then consider that limiting value as the "real" average of the thing that we measured. This "real" average is called the *population mean*.

1. Non-Fractal

For non-fractal objects, as more data is included, the averages of the data reach a limiting value which we therefore consider to be the "real" average.

(That is, the sample means converge to a finite, nonzero, limiting value that we identify as the population mean.)

2. Fractal

However, for a fractal, as more data are included, the averages of the data will continue to increase or continue to decrease. Thus there is no limiting value, we can consider to be the "real" average. The average does not exist.

(That is, the sample means do not converge to a finite, nonzero, limiting value, and thus there is no value that we can identify as the population mean.)

Non-Fractal

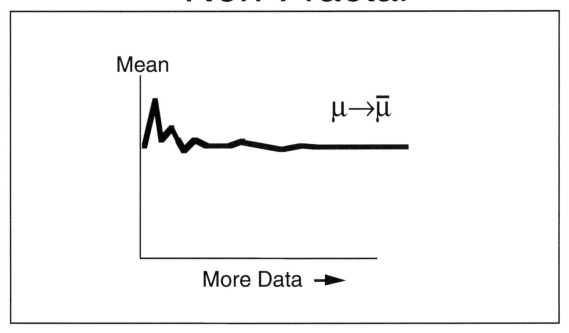

Fractal

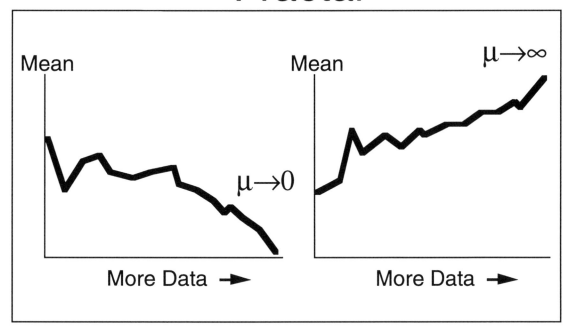

Self-Similarity Implies That the Moments Do Not Exist

Self-similarity means that there are an ever larger number of smaller pieces that resemble the larger pieces. Thus there are a few big pieces, many medium sized pieces, and a huge number of small pieces.

When we determine the average of the values of the sizes of the pieces, the most significant contribution to the average will come from either: (1) the frequent number of small values, in which case the average decreases as more data are included, or (2) the infrequent large values, in which case the average increases as more data are included.

A **moment** is the average of a property raised to a power.

The **mean** is the first moment. It is the average of the values of the sizes of the pieces. Self-similarity means that the mean will increase or decrease as more data are analyzed, depending on the relative contribution of the frequent small values versus the infrequent large values.

The **variance** is the second moment. It is the average of the square of the difference between each value and the mean value of the sizes of the pieces. Typically, self-similarity means that the variance will increase as more data are analyzed because there are an ever larger number of self-similar fluctuations.

We are so used to the properties of Gaussian distributions, where the mean and variance exist, that we assume that these moments must always exist. The existence of the mean and variance depend on the data satisfying the mathematical assumptions that these moments approach finite, nonzero limiting values as more data is analyzed. This fact is not well known. When the data are fractal, it does not satisfy these assumptions, the mean and variance do not exist, and therefore they are not useful to characterize the properties of fractal data.

The Average Depends on the Amount of Data Analyzed

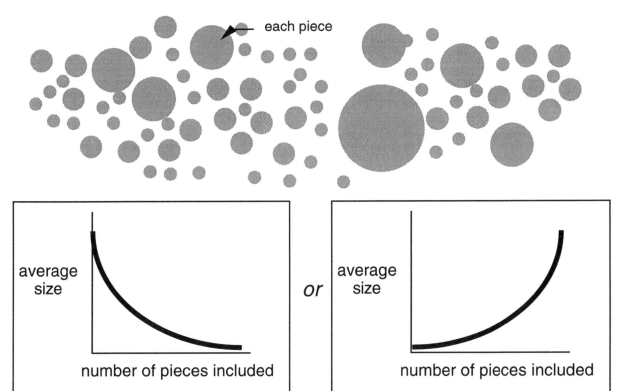

each piece

average size | *or* | average size
number of pieces included | | number of pieces included

$$\text{mean} = \frac{f_1x_1 + f_2x_2 + f_3x_3 + \ldots + f_nx_n}{f_1 + f_2 + f_1 + \ldots + f_n} \qquad f \text{ is the number of pieces of size } x$$

$$\lim_{x_i \to 0} \text{mean} = \lim_{x_i \to 0} \frac{\mathbf{f_1x_1} + f_2x_2 + f_3x_3 + \ldots + f_nx_n}{f_1 + f_2 + f_1 + \ldots + f_n} = 0$$

Contributions to the mean dominated by the <u>many</u> f_1 of the <u>smallest</u> sizes x_1.

$$\lim_{x_i \to \infty} \text{mean} = \lim_{x_i \to \infty} \frac{f_1x_1 + f_2x_2 + f_3x_3 + \ldots + \mathbf{f_nx_n}}{f_1 + f_2 + f_1 + \ldots + f_n} = \infty$$

Contributions to the mean dominated by the <u>few</u> f_n of the <u>biggest</u> sizes x_n.

Example Where the Average Does Not Exist:
The St. Petersburg Game

1. The Average Winnings is 50¢ for a Non-Fractal, Ordinary Coin Toss Game

Play the following game. Toss a coin. If it lands Heads, you win $1; if it lands Tails, you win nothing. The average payback to you is the sum of the probability of each outcome multiplied by the winnings associated with it, namely $(1/2) \times 0 + (1/2) \times 1 = \$0.50$. The more times N that you play this game, the closer the payback to you, averaged over all the N games, approaches $0.50. The house should be ready to pay out $0.50, on average, and you should be willing to bet $0.50 to play each game.

2. There is No Average Winnings for the Fractal, St. Petersburg Game

Now play a game formulated about 250 years ago by Niklaus Bernoulli and analyzed and published by his cousin Daniel Bernoulli. You toss a coin and continue to do so *until* it lands Heads. You get $2 from the house if it lands Heads on the first toss, $4 if on the second toss, $8 if on the third toss, $16, if on the fourth toss, and so on, so that with each additional toss the number of dollars that the house must pay is doubled. The average payback to you is $(1/2) \times 2 + (1/4) \times 4 + (1/8) \times 8 + (1/16) \times 16 \ldots = 1 + 1 + 1 + 1 \ldots = \infty$. There is NO number that we can identify as the average winnings!

The more times N that you play this game, the larger the payback to you averaged over all the N games. The average winnings after N games continues to increase as N increases. It does not reach a finite, limiting value that we can identify as the average winnings. The average winnings for playing this game does not exist. This game is fractal because the distribution of winnings (how much you win for each probability of winning) has a power law scaling.

This game is called the St. Petersburg paradox. Since there is a 50/50 chance that you will $2 on each game, you would be willing to put up $2 to play. But since the average winnings is infinite, the house would want you to put up more than all the money in the world to play. This type of game and the statistical properties associated with it were studied by mathematicians since it was first proposed. However, these ideas became separated from the main thread of probability theory that became popular among natural scientists.

Non-Fractal

$$\mu \rightarrow 1/2$$

Ordinary Coin Toss

Toss a coin. If it is tails win $0, if it is heads win $1.
The average winnings are: $2^{-1} \cdot 1 \ = \ 0.5$

Fractal

$$\mu \rightarrow \infty$$

St. Petersburg Game (Daniel Bernoulli)

Feller 1968 An Introduction to Probability Theory and Its Applications, vol. 1, Wiley, pp. 251-3.

Toss a coin. If it is heads win $2, if not, keep tossing it <u>until</u> it falls heads.

If this occurs on the N-th toss we win $ 2^N.

With probability 2^{-N} we win $ 2^N.

H	$2
TH	$4
TTH	$8
TTTH	$16 etc.

The average winnings are: $2^{-1} 2^1 \ + \ 2^{-2} 2^2 \ + \ 2^{-3} 2^3 \ + \ \ldots \ =$
$$1 \ + \ 1 \ + \ 1 \ + \ \ldots \ = \ \infty$$

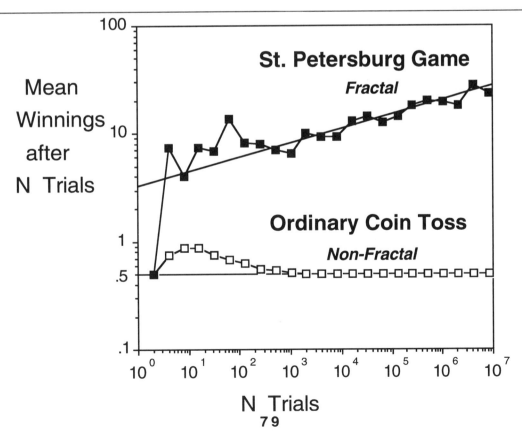

Mean
Winnings
after
N Trials

St. Petersburg Game
Fractal

Ordinary Coin Toss
Non-Fractal

N Trials

Example Where the Average Does Not Exist:
Diffusion Limited Aggregation (DLA)

1. The Average Density of a Non-Fractal Checkerboard

We want to determine the average density of the pixels in a black and white checkerboard. To find the average density within a circle of radius r, we count the number of black pixels within the circle and divide that by the total number of pixels. We repeat this measurement for circles of larger radii. As the radius increases, we find that there are some fluctuations in these averages. However, we also find that as the radius increases, these average densities approach a finite, limiting value that we therefore identify as the "real" average density of the checkerboard.

2. There is No Average Density of a Fractal DLA

We now want to determine the average density of a fractal object called a diffusion limited aggregate (DLA). It is formed by particles that are released one at a time from far away and randomly walk until they hit and stick to the growing structure. The DLA is self-similar. It has little spaces between the small arms and bigger spaces between the larger arms. We measure the average density within a circle of radius r. As the radius of the circle increases, we catch more of the ever bigger spaces between the ever larger arms. Thus, as the radius r of the circle increases, the average density within radius r decreases. The average density does not reach a finite, limiting value. It approaches zero. Therefore, the "real" average density for the DLA does not exist.

Non-Fractal

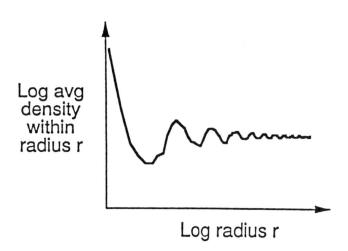

Fractal

Meakin 1986 In *On Growth and Form: Fractal and Non-Fractal Patterns in Physics*
Ed. Stanley & Ostrowsky, Martinus Nijoff Pub., pp. 111-135.

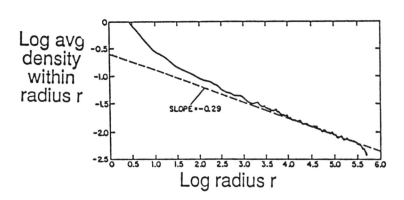

Example Where the Variance Does Not Exist:
The Roughness of Rocks

The edges of rocks are self-similar. Small sharp peaks over short distances are reproduced as ever larger peaks over ever larger distances.

The root mean square (RMS) of the heights along the rock profile is the square root of the average of the square of the heights along the rock. As the root mean square of the heights is measured over longer distances on the rock, ever larger peaks are included. Thus the root mean square of the heights increases with the distance measured along the rock. It does not approach a finite, limiting value.

The root mean square (RMS) of the heights along the rock is related to the variance of the fluctuations in height along the edge of the rock.

For a non-fractal object, as the variance is measured over longer distances, it approaches a finite value. Thus the variance exists and we identify that limiting value as the "real" value of the variance.

However, for the fractal edges of rocks the variance increases with the distance measured. The variance does not approach a finite value. Thus the variance does not exist.

Fractal Edges of Rocks

Brown and Scholz   1985   J. Geophys. Res. 90:12,575-12,582.

Height

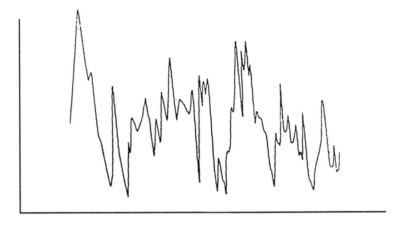

Distance

RMS
Height
(μ)

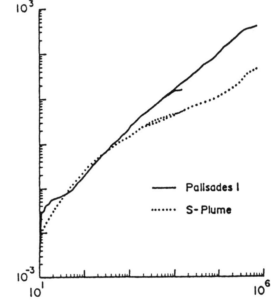

$$\sigma \rightarrow \infty$$

Palisades I

S-Plume

Profile Length (μ)

Example Where the Variance Does Not Exist:
Electrical Activity in Nerve Cells

Transient changes in the electrical voltage across the cell membrane, called action potentials, travel along the surface of nerve cells. Information is passed down the nerve axon in the timing between the action potentials.

Teich et al. analyzed the timing of the action potentials from nerves in the ear connected to the auditory cells that encode information about sound and from nerves connected to the vestibular cells that encode information about acceleration that is used for balance. They divided their records into consecutive time windows and counted the number of action potentials in each window. They determined how the number of action potentials in each time window depended on the length of the time windows. The time resolution was set by the length of the time windows. Thus they determined how the properties of the timing of the action potentials depended on the time resolution used to measure them.

For example, they studied the number of action potentials per second in each time window, which is called the *firing rate*. They evaluated the firing rate for time windows of different lengths. For a non-fractal, the fluctuations in the firing rate would be less in longer windows, because the statistical inaccuracies would average out. Instead, for the auditory nerve cells involved in hearing, they found that the variation in the firing rate decreased more slowly than for a non-fractal as the window length increased.

This result is surprising if you only know about the statistical properties of non-fractal distributions. For those distributions the fluctuations decrease at a certain rate as more data are included.

However, for fractal distributions, the fluctuations decrease more slowly as more data are included. Thus their result implies that the firing rate is fractal. This means that the action potentials are correlated with each other. These small correlations become more significant when the action potentials are analyzed over longer times in the longer time windows.

Electrical Activity of Auditory Nerve Cells
Teich, Johnson, Kumar, and Turcott 1990 Hearing Res. 46:41-52

Data:

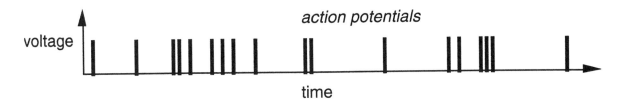

voltage

action potentials

time

Divide the record into time windows:

Count the number of action potentials in each window:

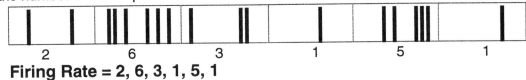

2   6   3   1   5   1

Firing Rate = 2, 6, 3, 1, 5, 1

Repeat for different lengths of time windows:

8   4   6

Firing Rate = 8, 4, 6

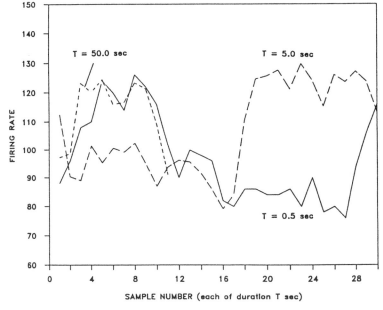

The variation in the firing rate decreases <u>slowly</u> at longer time windows.

Statistical Analysis of the Electrical Activity in Nerve Cells

1. Pulse Number Distribution

Teich et al. divided their records of action potentials recorded from nerves in the ear into consecutive time windows. They counted the number of action potentials in each time window. They evaluated how often each number of action potentials occurred in these time windows. This is called the **pulse number distribution**.

For a non-fractal process, as the window size is increased, the pulse number distribution becomes smoother and more like a Gaussian distribution. This property is called the Central Limit Theorem. This was the case for the data from the vestibular cells involved in balance.

This was not the case for the data from the auditory cells involved in hearing. Those distributions became rougher as the length of the time window was increased. This roughness arises from correlations in the times between the action potentials.

The proof of the Central Limit Theorem requires that the variance exist and that it has a finite value. For a fractal, the variance is not defined, and thus the Central Limit Theorem does not apply. As more data are included fractal distributions do not become smoother or more Gaussian. Rather, the distributions become rougher because the correlations that link the deviations together in a self-similar way become more noticeable over longer times. Teich et al. wrote that "the irregular nature of the long count pulse number distributions does not arise from statistical inaccuracies associated with insufficient data, but rather from event clustering inherent in the auditory neural spike train."

2. Fano Factor

The **Fano factor** is equal to the variance divided by the mean of the number of action potentials in the time windows. For a non-fractal process where the time between the action potentials is random, the pulse number distribution is a Poisson distribution, and the Fano factor is 1. They found that the Fano factor increased with the length T of the time windows used to measure it. The Fano factor F had a power law scaling relationship that F was proportional to T^d where d is the scaling exponent.

Statistical Analysis of Action Potentials
Teich 1989 IEEE Trans. Biomed. Engr. 36:150-160

Counts in the windows are: 2, 4, 3, 1, 1, 1

Determine the **Mean** and **Variance** of the counts.

Determine the **Pulse Number Distribution**:
Number (#) of windows of size T with n counts.

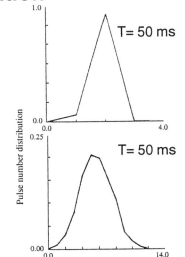

Pulse Number Distribution

Non-Fractal Vestibular Neuron	as, T ↑ the PND becomes <u>more</u> Gaussian	
Fractal Auditory Neuron	as, T ↑ the PND becomes <u>less</u> Gaussian	

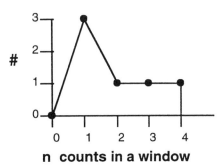

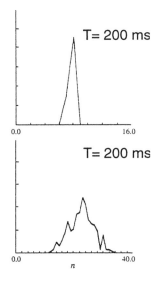

Fano Factor = Variance/Mean

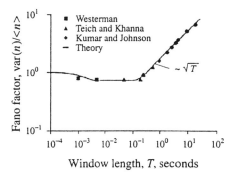

The Fano Factor = Variance/Mean increases with the window length T for the auditory neurons.

Example Where the Variance Does Not Exist: Blood Flow in the Heart

Bassingthwaighte et al. measured blood flow in the heart by adding radioactive tracers to the blood. These tracers leak out of the blood vessels and become trapped in the muscle of the heart. The amount of tracer that leaks out is proportional to the blood flow in that region. Thus they determined the blood flow by measuring the amount of radioactivity in different pieces of the heart.

They determined the **relative dispersion**, RD, which is equal to the standard deviation divided by the average of the blood flow measured in a piece of weight w. They measured the blood flow in pieces of different weight. The weight of the pieces set the spatial resolution of the measurement. They found that the relative dispersion had the power law scaling relationship that RD was proportional to w^{1-d}, where d is the fractal dimension.

Thus blood does not flow evenly throughout the heart. The flow of blood through the heart is fractal. There are regions of higher than average blood flow and regions of lower than average blood flow. The pattern of blood flow is self-similar. There are ever smaller regions of higher and lower than average blood flow. The dimension d is a measure of the correlation between regions with different amounts of blood flow.

They showed that if the large blood vessels connect to slightly unequal smaller branches, then the relative dispersion would increase after each branching. This would explain why the relative dispersion increases as it is measured from pieces of smaller weight that correspond to the finer spatial resolution.

Blood Flow in the Heart

Bassingthwaighte and van Beek 1988 Proc. IEEE 76:693-699

When the blood flow is measured using <u>smaller</u> pieces, the relative dispersion <u>increases</u>.

Data

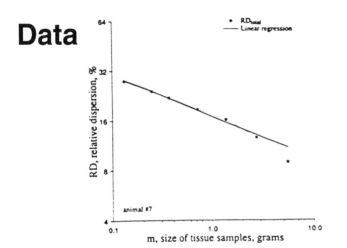

$$RD = \frac{\text{standard deviation}}{\text{mean}}$$

$$RD \propto m^{1-d}$$

$$d = 1.2$$

Model

$a = .5 + \varepsilon$

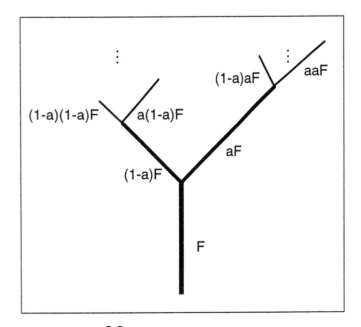

Example Where the Variance Does Not Exist:
Volume of Breaths

Hoop et al. measured the volume of air in each breath of a rat. We used the **rescaled range analysis** developed by **Hurst** to analyze the values of the volumes of consecutive breaths. The rescaled range **R/S** is a measure of the variance. We evaluated the rescaled range over time windows of different length T. The length T of the time windows is called the **lag**. The values of the lag set the temporal resolution. Thus we determined how the rescaled range depended on the temporal resolution at which it was measured.

A fractal set of values has small variations in the values over brief times that look similar to larger variations in the values over longer times. All the past values affect the future values. The rescaled range R/S of a fractal time series will have the power law scaling relationship that R/S is proportional to T^H, where H is called the **Hurst exponent**.

When H = 1/2, the differences between consecutive values are uncorrelated. When 1/2 < H < 1, the differences between consecutive values are said to be **persistent**. This means that increases at one time are more likely to be followed by increases at all later times, and decreases at one time are more likely to be followed by decreases at all later times. When 0 < H < 1/2, the differences between consecutive values are said to be **anti-persistent**. This means that increases at one time are more likely to be followed by decreases at all later times, and decreases at one time are more likely to be followed by increases at all later times.

We found that the data were a straight line on a plot of Log (R/S) versus Log (T). This means that the rescaled range R/S of the volumes of the breaths was proportional to T^H. Thus this power law scaling indicates that the sequence of volumes of these breaths is fractal. The Hurst exponent H was greater than 1/2. This means that there are persistent correlations in the volumes of the sequence of breaths. The discovery of the fractal nature of these volumes and their positive correlations may provide information about the feedback mechanisms that control the amount of air in each breath.

90

Volume of Consecutive Breaths
Hoop, Kazemi, and Liebovitch 1993 Chaos 3:27-29

Hurst Rescaled Range Analysis

R = range of the deviation of the running sum from the mean over a time window of length T

S = standard deviation over a time window of length T

R / S = rescaled range

T = time window in which R / S is measured

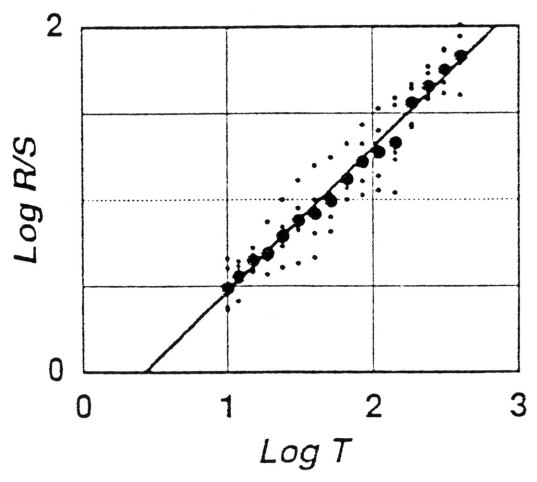

Example Where the Variance Does Not Exist: Evolution

In the beginning of the 20th century, it was not known whether: (1) mutations occur *all the time* but are only selected when there is a change in the environment or (2) mutations occur *only* in response to a change in the environment.

Luria and Delbruck let a cell multiply many times and then challenged its descendants with a killer virus. They ran many such experiments and determined how many cells were resistant to the virus at the end of each experiment.

If mutations occur all the time, then by chance, some cells will become resistant to the killer virus *before* it is given. This resistant cell will divide and give rise to resistant daughter cells in subsequent generations. If the resistant cell is produced early on, it will form many resistant daughters in the many subsequent generations. If the resistant cell is produced later on, it will not have time to form as many resistant daughters. Each time the experiment is run the mutations will occur at different times. The variations in the timing of the appearance of the resistant mutations are amplified by the number of resistant daughters they produce in the subsequent generations. This results in a large variation in the final number of resistant cells when the experiment is repeated many times.

If the mutations occur only in response to the virus, then by chance, some cells will become resistant to the killer virus *when* it is given. However, in this case, they will not have time to form resistant daughters. Thus there will be only a small variation in the final number of resistant cells when the experiment is repeated many times.

Luria and Delbruck found that there was a large variation in the number of resistant cells. Thus they concluded that random mutations occur all the time, but are only selected when there is a change in the environment.

Is Evolution an **Editor** or a **Composer** ?

Luria & Delbruck 1943 Genetics 28:491-511
(Cairns, Overbaugh & Miller 1988 Nature 335:142-145; Levin, Gordon & Stewart 1989 preprint)

Experiment:

Let the colony grow and then challenge it with a killer virus.
Determine the number of mutant cells at the end of each experiment.

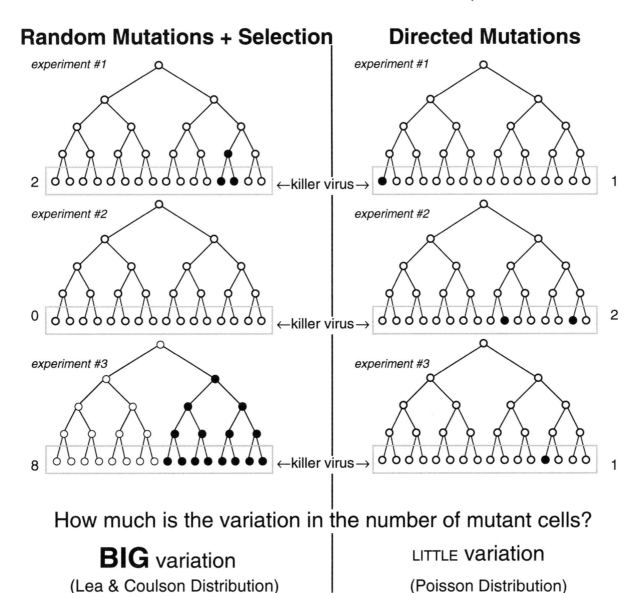

Random Mutations + Selection | **Directed Mutations**

How much is the variation in the number of mutant cells?

BIG variation

(Lea & Coulson Distribution)

LITTLE variation

(Poisson Distribution)

The Distribution of Mutations Is the Same as in the St. Petersburg Game

If we run many experiments where mutations occur all the time, what is the average number of mutant cells at the end of the experiments?

Each cell has the same probability of giving birth to a mutant cell. There are few cells at the beginning of the experiment. Thus there is a low probability that a mutation will occur in the first generation of cells. However, if a mutation does occur in the first generation, that cell will produce 2^N daughter cells in the N generations until the end of the experiment. In the second generation, there are already twice as many cells as the first generation. Thus the probability that a mutation occurs in the second generation is twice as great as that in the first generation. However, that mutant cell will produce only 2^{N-1} daughters in the remaining $N-1$ generations. Similarly, for all subsequent generations.

The total number of mutant cells at the end of the experiment is equal to the probability of a mutation occurring in each generation multiplied by the number of daughter cells that it produces until the end of the experiment. *This means, on average, that each generation contributes the same number of mutant cells to the final number of mutant cells.* This is the same calculation as determining the winnings in the St. Petersburg game described previously. Each generation corresponds to one play of the game, the probability of winning on that play corresponds to the probability that a mutation occurs in that generation, and the money won corresponds to the number of resistant daughter cells at the end of the experiment,

As is true for the average winnings of the St. Petersburg game, the average number of mutant cells at the end of the experiment is not defined. The average found for a number of experiments will increase as the number of experiments is increased. Because of this large variability, it is not known how to compare the average number of final mutant cells from two sets of experiments to determine if the different conditions of the experiments affected the mutation rate.

Random Mutations

Lea & Coulson 1949 J. Genetics 49:264-285.

Mandelbrot 1974 J. Appl. Prob. 11:437-444.

If the probability of a mutation per cell $= \dfrac{1}{16} = 2^{-4}$

FOR EACH GENERATION:

Probability per cell of one Mutation	X	No. of Cells in this gener- ation	=	Prob. of one Mutation in this generation	X	No. of Offspring at the end	=	Expected Number of Mutants at the end from this generation

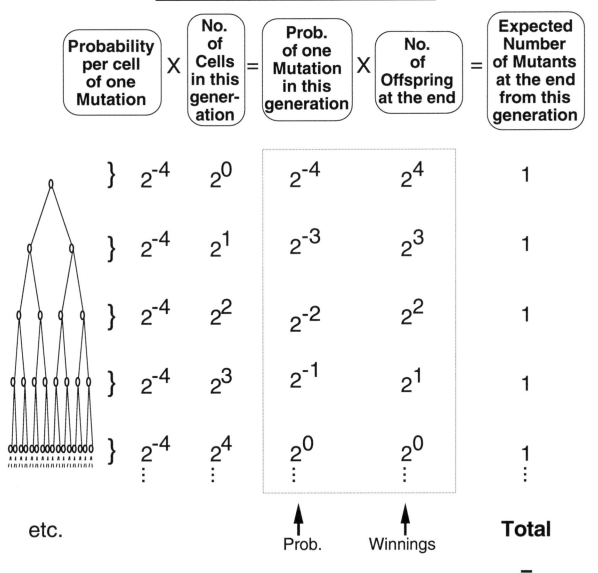

}	2^{-4}	2^{0}	2^{-4}	2^{4}		1
}	2^{-4}	2^{1}	2^{-3}	2^{3}		1
}	2^{-4}	2^{2}	2^{-2}	2^{2}		1
}	2^{-4}	2^{3}	2^{-1}	2^{1}		1
}	2^{-4}	2^{4}	2^{0}	2^{0}		1
		⋮	⋮	⋮		⋮

etc.

↑ Prob.   ↑ Winnings   **Total**

=

St. Petersburg Game ∞

Statistical Properties and the Power Spectra

A sequence of values measured in time can be broken down into a series of periodic oscillations of different frequencies. A plot of the amount present in each frequency is called the **power spectrum**.

The power spectra of fractal objects in space or fractal processes in time reflect the self-similarity, scaling relationship, and statistical properties of fractals.

Low frequencies in the power spectrum correspond to coarse resolution. High frequencies in the power spectrum correspond to fine resolution. Thus the power spectrum is a measure of the amount present in structures of different sizes.

Self-similarity means that there is a relationship between the power at high frequencies (fine resolution) and the power at low frequencies (coarse resolution). This relationship has a **power law scaling** that the energy at a given frequency is proportional to $1/f^{\alpha}$. This form is called **1/f** ("one-over-f") **noise**, even when the exponent α is not equal to 1.

Depending on the value of α, the **total power** at all the frequencies in the power spectrum and the **average power do not exist**. When $\alpha \geq 1$, then the total energy increases as the lowest frequency used to measure it decreases. That is, the longer the interval of data that is analyzed, the larger the total power in the power spectrum. When $\alpha \leq 1$, then the total power increases as the highest frequency used to measure it increases. In this case, the ever shorter intervals of data contain ever larger amounts of power.

An example of 1/f noise in time is the electrical signal generated by the contraction of the heart. An example of 1/f noise in space is the spatial distribution of a radioactive tracer in the liver.

Fractal Power Spectra P(f)

If the variance $\rightarrow \infty$, then there are fluctuations at <u>ALL</u> scales, and $P(f) = \frac{1}{f^{\alpha}}$

in time

electrical activity when the heart contracts
Goldberger, Bhargava, West, and Mandell 1985 Biophys. J. 48:525-528

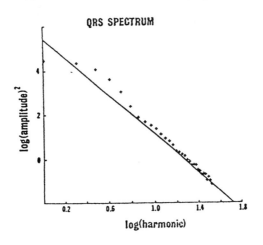

in space

radioactive isotope distribution in the liver
Cargill, Barrett, Fiete, Ker, Patton, and Seeley 1988 SPIE 914 Medical Imaging II, pp. 355-361.

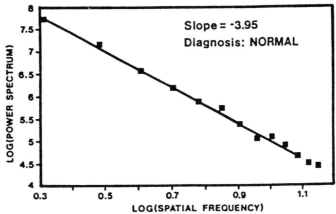

How to Measure the Properties of Fractal Data

The statistical properties, such as the average or variance, of a fractal object or process depend on the resolution used to measure them. Thus **it does little good to measure these statistical properties at only one resolution**.

The meaningful measurement is **to determine how these statistical properties depend on the resolution used to measure them**.

Usually, the scaling relationship for a statistical property $Q(r)$ will have the power law form that $Q(r)$ is proportional to r^b, where r is the resolution. The exponent b is related to the fractal dimension.

A number of different statistical properties $Q(r)$ have been used to analyze fractal data, including the:

1. mean (average);

2. variance;

3. standard deviation;

4. relative dispersion (standard deviation/variance);

5. Fano factor (variance/mean);

6. mean squared deviation; and

7. rescaled range (range of the running sum of the deviations from the mean divided by the standard deviation).

When the moments, such as the mean and variance, don't exist, what should I measure?

You should measure how a property Q(r) depends on the resolution r used to measure it.

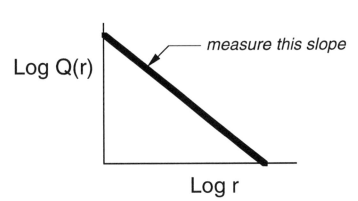

examples of Q(r):

Mean

e.g. average density within a circle as a function of the radius of that circle

Meakin 1986 In On Growth and Form, ed. Stanley & Ostrowksy Nijhoff. pp. 111-135

Relative Dispersion (standard deviation / mean)

e.g. relative dispersion of blood flow as a function of the mass of the tissue sample

Bassingthwaighte and van Beek 1988 Proc. IEEE 76:693-699

Fano Factor (variance / mean)

e.g. Fano factor of the number of action potentials within time windows as a function of the length of time windows

Teich, Johnson, Kumar, and Turcott 1990 Hearing Res. 46:41-52

Mean Squared Deviation

e.g. mean squared deviation of a walk generated from the base pair sequence in DNA as a function of the length along the DNA

Peng et al. 1992 Nature 356:168-170.

Hurst Rescaled Range

e.g. maximum minus the minimum value of the running sum of the deviations from the mean normalized by the standard deviation of volumes of breaths measured within a time window as a function of the length of the time window

Hoop et al. 1993 Chaos 3:27-29.

More Examples of the Statistical Properties
of Fractals

We have already seen that the statistical properties of fractals are present in the timing of the action potentials recorded from nerves that carry information about sound that we use for hearing, in the spatial distribution of blood flow in the muscle of the heart, in the sequence of the volumes of breaths, in the number of mutated cells produced by Darwinian evolution, in the timing of the electrical activity of the heartbeat, and in the spatial distribution of radioactive tracer in the liver.

Additional examples of these statistical properties include the changing electrical voltage across the cell membrane of white blood cells (T-lymphocytes), the sequence of base pairs in DNA where each base pair is assigned a number and the running sum of these numbers is analyzed, and the duration in time of consecutive breaths.

Biological Examples of the Statistical Properties of Fractals

action potentials in nerve cells

blood flow in the heart

volumes of consecutive breaths

mutations

electrical activity of the heartbeat

distribution of tracer in the liver

membrane voltage of T-lymphocytes

base pair sequence in DNA

durations of consecutive breaths

Biological Implications of the Statistical Properties of Fractals

1. Not Gaussian or Asymptotically Gaussian

The statistical knowledge of most scientists is limited to the statistical properties of **Gaussian** distributions. **Fractals do not have the properties of Gaussian distributions.** In order to understand the many fractal objects and processes in the natural world, it is required to learn about the properties of **stable** distributions. Stable distributions are more general than Gaussian distributions.

2. The Average and Variance Do Not Exist

The moments of a fractal, such as the mean and variance, do not exist. As more data are included, the measurements of these moments do not approach finite, limiting values.

3. Large Variations

The variance of a fractal increases as more data are analyzed. The average values measured for the properties of the data will have wide variation from one time to another and among repetitions of the same experiment.

4. When Are These Large Variations Significant?

When the variance of a fractal increases as more data are analyzed, we do not know how to perform statistical tests to determine if the parameters of the mechanism that generated the data have changed from one time to another or between experiments run under different conditions.

The statistical tests taught in the usual statistics courses are based on the assumption that the variance is finite. These tests are not valid to analyze fractal data where the variance is infinite. **It would be very worthwhile for mathematicians to formulate statistical tests for fractal distributions where the variance is infinite.**

Statistical Properties

What they taught you in school	Fractals

	Stable Distributions
moments → nonzero, finite	*moments → 0, ∞*
σ → finite	*σ → ∞*
statistical tests to tell if parameters differ at different times or between different experimental conditions:	*statistical tests to tell if parameters differ at different times or between different experimental conditions:*
t, F, ANOVA non-parametric	?

Biological Implications of the Statistical Properties of Fractals (continued)

5. "Nonstationary" Does Not Mean That Things Are Changing

When the measurements of the moments keep changing as more data are included, then a process is said to be "**nonstationary**." This is a poor choice of a word, because it suggests that the process is changing in time. This is not necessarily true. **A generating mechanism whose operation is fixed in time can produce a fractal output whose moments keep changing in time.** When the moments of the data are "nonstationary," this does not necessarily mean that the mechanism that produced the data is changing in time.

6. When the Variance Looks Big, Maybe It's Infinite

Sometimes there is a large variability in a property measured in repetitions of an experiment. For example, such variability is present in the number of mutant cells in evolution and in the time durations between the steps of growth and division (cell cycle time) in cancer cells. Scientists analyzing this kind of data who are familiar only with Gaussian statistics usually assume that the data were produced by a process with a finite variance. Any time such data with large variability are found, it may be worthwhile to determine if the variance does, or does not, have a finite, limiting value. This can be done by measuring how the variance depends on the amount of data included. If the variance increases with the amount of data included, then the data have fractal properties and the variance does not exist.

7. Importance of the Scaling Relationship

When the moments, such as the mean and variance, do not exist, their values will depend on the resolution used to measure them. Thus the **measurement of the moments at one resolution is not meaningful**. What is meaningful is to determine **how the moments depend on the resolution used to measure them**. The form of this dependency is called the scaling relationship, and it is characterized by the parameter called the fractal dimension.

More Statistical Lessons

nonstationary

This word means that the moments do not exist.

(The moments do not reach finite, limiting values.)

It does **not** mean that the mechanism that produced the data is changing in time.

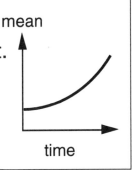

BIG variance?

Check its limiting value.

Maybe, it's ∞.

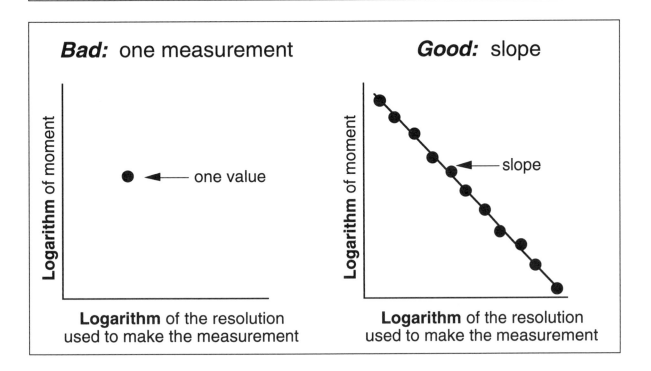

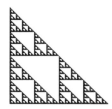

FRACTALS
Summary

Summary of Fractals

1. Fractals Are Self-Similar

Self-similarity means that ever smaller pieces of an object resemble the larger pieces of an object.

2. Self-Similarity Produces a Scaling Relationship

When the value of a property is measured at finer resolution, it will include ever more of the smaller, self-similar pieces. Thus the value measured for a property depends on the resolution used to make the measurement. This is called the **scaling relationship**.

The scaling relationship usually has the *power law* form that the value $Q(r)$ measured for a property at resolution r is proportional to r^b.

3. The Dimension is a Measure of Self-Similarity and Scaling

The dimension is a quantitative measure of the self-similarity and the scaling. The **dimension** tells us how many new pieces are found when an object is examined at finer resolution.

There are many different types of dimension. The **fractal dimension** characterizes the space filling properties of an object. The different types of fractal dimension share the common feature that the number of pieces $N(r)$ found at resolution r is proportional to r^{-d}, where d is the fractal dimension.

4. Fractals Have Surprising Statistical Properties

The statistical properties of fractals are described by **stable distributions**. Stable distributions are more general than the Gaussian distributions they taught you in school. The moments, such as the average and variance, of stable distributions may not have nonzero, finite values.

Summary of Fractal Properties

Self-Similarity
Pieces resemble the whole.

Scaling
The value measured depends on the resolution.

Dimension
How many new pieces are found as the resolution is increased.

Statistical Properties
Moments may be zero or infinite.

Where to Learn More about Fractals

There are many good books on fractals and hundreds of research articles published in journals each year. Some references, at different levels, that can lead you further into the mathematical details and the applications of fractals are the following:

1. Mandelbrot

The Fractal Geometry of Nature and its earlier edition *Fractals: Form, Chance, and Dimension* by Mandelbrot made the scientific community aware of fractals. Mandelbrot developed some of the mathematics of fractals and applied them to many different scientific fields. His book is beautiful, passionate, and at a mathematical level that is sometimes frustratingly too simple and too complex at the same time.

2. Fractals in Mathematics

The mathematics of fractals is based on measure theory and topology. Two books that provide a clear and rigorous introduction to the mathematics of fractals are *Fractals Everywhere* by Barnsley and *Measure, Topology, and Fractal Geometry* by Edgar.

3. Fractals in Physics and Chemistry

Fractals by Feder and *The Fractal Approach to Heterogeneous Chemistry* edited by Avnir provide an introduction to fractals at the mathematical level of the calculus and review applications of fractals in physics and chemistry.

4. Fractals in Biomedical Research

Fractal Physiology by Bassingthwaighte, Liebovitch, and West provides an introduction to fractal properties both at a qualitative level and at the mathematical level of elementary calculus. It gives detailed descriptions and references to many biomedical applications of fractals. Additional biomedical applications are described in *Fractal Geometry in Biological Systems*, edited by Iannaccone and Khokha.

110

Books About Fractals

classic

> ## B. B. Mandelbrot
> ### *The Fractal Geometry of Nature*   1983   W. H. Freeman

mathematics

> ## G. A. Edgar
> ### *Measure, Topology, and Fractal Geometry*
> 1990   Springer-Verlag
>
> ## M. Barnsley
> ### *Fractals Everywhere*   1988   Academic Press

physics & chemistry

> ## J. Feder
> ### *Fractals*   1988   Plenum
>
> ## D. Avnir
> ### *The Fractal Approach to Heterogeneous Chemistry*
> 1989   John Wiley & Sons

biomedical

> ## J. Bassingthwaighte, L. Liebovitch, & B. West
> ### *Fractal Physiology* 1994   Oxford Univ. Press
>
> ## P. M. Iannaccone & M. Khokha
> ### *Fractal Geometry in Biological Systems* 1996   CRC Press

Part II
CHAOS

Deterministic systems with output so
complex that it mimics random behavior.

$$X(t+\Delta t) = 3.95 \, X(t) \, [1-X(t)]$$

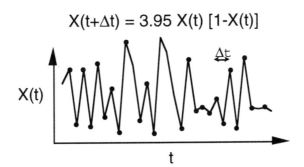

CHAOS

Introduction

Two Sets of Data That Look Alike

Each graph on the facing page consists of a set of data. Each data value is measured at one instant in time. These instants in time are separated by constant intervals in time. The n-th measurement is denoted by x(n). The values measured are shown as dots. Consecutive values have been connected by straight lines.

Both sets of data also have the **same statistical properties**. That is, both sets of data have the same average, variance, and power spectrum.

Each graph does not have the same values of x(n). However, the pattern of variation in each graph looks similar. The values of x(n) seem to vary in a **random** way with n. Therefore, it seems reasonable to believe that both sets of data were generated by an inherently random mechanism.

1. Left: Data Set #1

The values x(n) seem to change randomly from one point in time to the next point in time.

2. Right: Data Set #2

The values x(n) seem to change randomly from one point in time to the next point in time.

116

Data 1

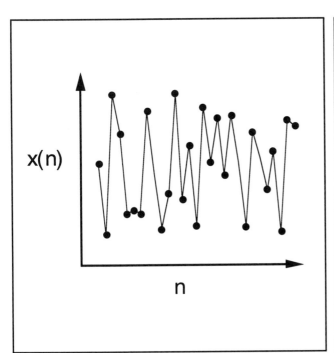

Data 2

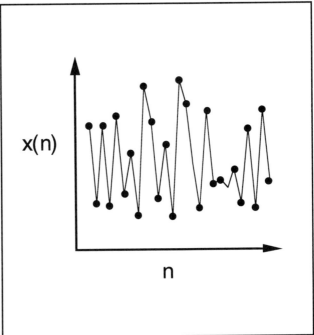

These two sets of data have the **same**:

mean

variance

power spectrum

The Difference between Randomness and Chaos

The pattern of x(n) in both data sets appears to have been generated by a random mechanism. However, not everything that looks random really is random.

One of these data sets was generated by a random mechanism. However, the other data set was not generated by a random mechanism at all. It was generated by a deterministic mechanism.

1. Left: Data Set #1 Was Generated by a Random Mechanism

This data set was generated by a **random** mechanism. Each new value of the data x(n) was chosen at random.

2. Right: Data Set #2 Was Generated by a Non-Random, Deterministic Mechanism

This data set was generated by a **deterministic** mechanism. Deterministic means that the next value of the data was computed from the previous values.

The next value of x(n+1) was computed from the previous value x(n) by using the simple rule that   **x(n+1) = 3.95 x(n) [1 - x(n)]**.

The phenomenon that a deterministic mechanism can generate data that looks as if it were generated by a random mechanism is called "**chaos**."

The word chaos was chosen to describe the complex output of these deterministic mechanisms. Chaos is a poor choice of a word for this phenomenon because here it means just the *opposite* of its common usage of "disordered." Here, ***chaos means that the output of a deterministic system is so complex that it mimics the output generated by a random mechanism. It does NOT mean that a system is driven by disorder, randomness, or chance.***

Data 1   Data 2

RANDOM   CHAOS

random   deterministic

$x(n) = RND$ $x(n+1) = 3.95 \, x(n) \, [1-x(n)]$

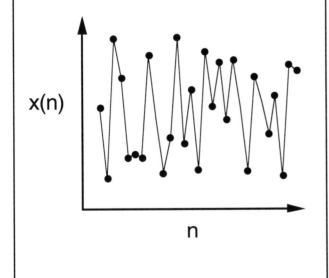

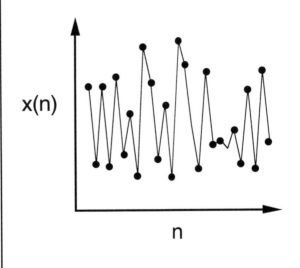

A Simple Equation Can Produce
Complicated Output

The phenomenon of chaos is surprising.

Let's follow the deterministic equation, step by step, to see how it produces a set of data that looks random:

The first value of x at n=1 was chosen to be x(1) = .892.

The value x(n+1) at the (n+1)-th step is computed from the previous value x(n) at the n-th step by using the equation that x(n+1) = 3.95 x(n) [1 - x(n)].

Thus, to compute the next value of x at n=2, multiply 3.95 by .892 and then multiply that result by (1-.892). This yields x(2) = .380.

To compute the next value of x at n=3, multiply 3.95 by .380 and then multiply that result by (1-.380). This yields x(3) = .931.

Continue the same process.

This deterministic mechanism alone produces a seemingly random sequence of values x(n). That is why the discoverers of this phenomenon called it **chaos**.

$$x(n+1) = 3.95[x(n)][1-x(n)]$$

$x(1) = .892$

$x(2) = 3.95[.892][1-.892] = .380$

$x(3) = 3.95[.380][1-.380] = .931$

$x(4) = 3.95[.931][1-.931] = .253$

$x(5) = 3.95[.253][1-.253] = .747$

$x(6) = 3.95[.747][1-.747] = .746$

etc.

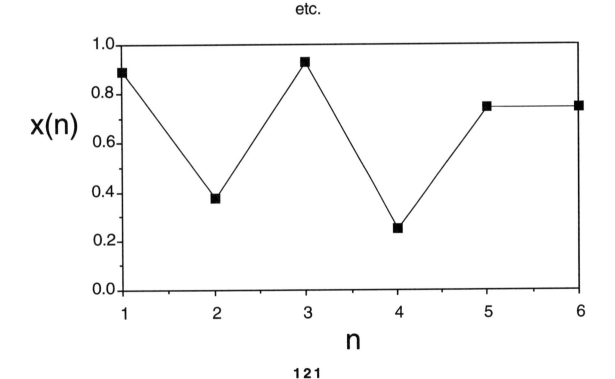

How to Tell Randomness from Chaos

If the data generated by random and deterministic mechanisms look so much alike, how can we tell which mechanism actually generated the data?

Although the sequence of values of these two data sets looks **similar**, a transformation of these data looks **different** for random and deterministic mechanisms. This transformation is a plot called a phase space and the points in it are called the **phase space set**. Thus the phase space set can be used to determine if the sequence of values was generated by a random or a deterministic mechanism.

The phase space set is constructed in the following way:
We take the first two values x(1) and x(2) and make believe that these two values are the X and Y coordinates of a point. That is, we set X = x(1) and Y = x(2), and plot a point at those coordinates.
Then we take the next two values x(2) and x(3), and again make believe that these two values are the X and Y coordinates of a point. Thus we now plot a point at coordinates X=x(2), Y=x(3).
Then we take the next two values x(3) and x(4), and again make believe that these two values are the X and Y coordinates of a point. Thus we now plot a point at coordinates X=x(3), Y=x(4).
And so on.

1. Left: Random

The phase space set of points produced from the random data **fills up** the 2-dimensional space.

2. Right: Deterministic Chaos

The phase space set of points from the chaotic data **does not fill up** the 2-dimensional space.   It forms a 1-dimensional object called a strange attractor.
The parabola seen in the phase space set reveals to us the equation x(n+1) = 3.95 x(n) [1 - x(n)] that was used to generate the data.

Data 1

Data 2

RANDOM

random

$$x(n) = RND$$

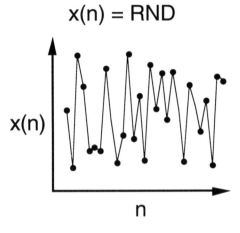

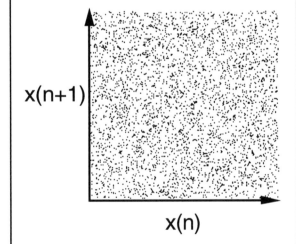

CHAOS

deterministic

$$x(n+1) = 3.95\ x(n)\ [1-x(n)]$$

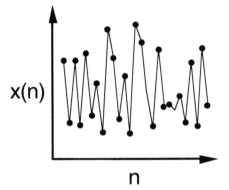

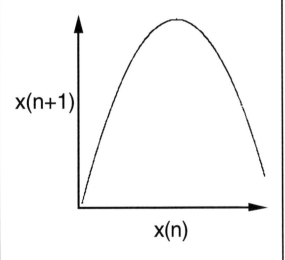

Definition of Chaos

Chaos: Complex output that mimics random behavior that is generated by a simple, deterministic system.

---

1. A Chaotic System Is Deterministic

The present values of the variables depend entirely on their recent, previous values. There is no element of chance. This is called a **dynamical system**. Some dynamical systems are chaotic and some are not chaotic.

Mathematically, dynamical systems are represented by difference or differential equations. These equations consist of a set of variables. The new values of the variables are computed from the previous values of the variables. In difference equations, time advances in discrete steps. The new values of the variables are computed from the values of the variables at earlier time steps. In differential equations, time advances continuously. The new values of the variables are computed from the values of the variables and their rate of change at earlier moments in time.

2. A Chaotic System Has Only a Small Number of Independent Variables

The equations that describe how the present values of the variables are computed from their previous values have only a small number of independent variables.

3. The Output of a Chaotic System Is Very Complex

The data that are generated by a chaotic system are so complex that they mimic the data generated by a system based on chance.

Chaos

Definition:

Deterministic

predict that value

these values

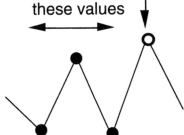

Small Number of Variables

$$x(n+1) = f(x(n), x(n-1), x(n-2))$$

Complex Output

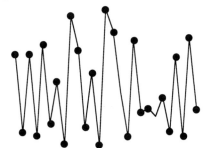

The Properties of Chaos

1. Phase Space

A sequence of values of data measured in time can be transformed into an object in space. This object is called the **phase space set**. Some properties are easier to determine by analyzing the phase space set rather than by analyzing the original data directly. For example, when the **fractal dimension** of the phase space set is **high**, then the data were generated by a **random** mechanism. When the fractal dimension of the phase space set is **low**, then the data were generated by a **deterministic** mechanism.

2. Sensitivity to Initial Conditions

If we rerun a non-chaotic system with almost the same starting values, we get almost the same values of the variables at the end. However, if we rerun a *chaotic* system with *almost the same starting values*, we get *very different* values of the variables at the end. This is called sensitivity to initial conditions. Chaotic systems **amplify small differences in initial conditions into large differences**. This is called the *Butterfly Effect*, because whether or not a butterfly beats its wings in Gliwice, Poland determines if there will be a thunderstorm a week later in Boca Raton, Florida.

3. Bifurcations

The behavior of a system can change abruptly with a small change in the value of a parameter. This is called a **bifurcation**.

4. Good News and Bad News About Analyzing Data

New methods now make it possible to analyze random-looking data to determine if they were produced by a random or a deterministic mechanism. However, there are practical difficulties in using these methods.

5. Control of Chaos

If the analysis shows that the data were produced by a deterministic mechanism, then we may be able understand and perhaps even control it.

Chaos

Properties:

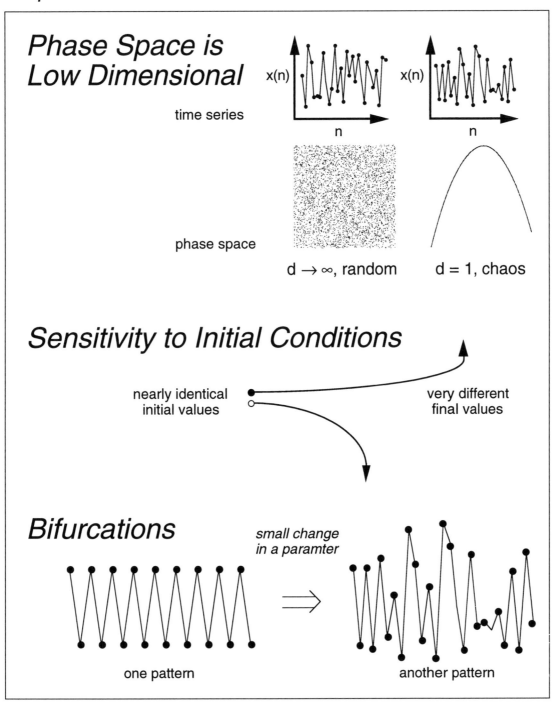

CHAOS
Phase Space

Phase Space

About a hundred years ago the mathematician Poincaré showed how to transform a sequence of values in time into an object in space. This transformation replaces an analysis in time with an analysis in space. The space is called the **phase space**. The object in the phase space is called the **phase space set**. The act of transforming the sequence of values in time into the object in space is called an **embedding**.

This transformation is useful because certain properties of the data are easier to determine from the phase space set than from the original sequence of values in time.

For example, let's follow the location of a baseball in Fenway Park in Boston. The sequence of values in time of its distance left, right, and above home plate are given by $X(t)$, $Y(t)$, and $Z(t)$. At a time t_1 the location of the baseball is given by the values X_1, Y_1, and Z_1. The point in a 3-dimensional phase space with coordinates $X=X_1$, $Y=Y_1$, and $Z=Z_1$, corresponds to the location of the baseball at that time. As the baseball moves, the values X, Y, and Z change. These values correspond to the coordinates of the point in the phase space. Thus the point moves in the phase space. This moving point traces out an object in the phase space. This object is the phase space set.

The sequence of values, $X(t)$, $Y(t)$, and $Z(t)$, is not limited to positions in space. For example, $X(t)$ could be the value of the temperature, $Y(t)$ the value of the the potassium concentration, and $Z(t)$ the value of the electrical voltage measured from a cell. At a time t_1, these measurements have values X_1, Y_1, and Z_1. The point in a 3-dimensional phase space with coordinates $X=X_1$, $Y=Y_1$, and $Z=Z_1$ corresponds to the state of the cell at that time. As the state of the cell changes, the values measured for X, Y, and Z change. These values correspond to the coordinates of the point in the phase space. As the state of the cell changes, these values change, and the point representing the cell moves in the phase space. As it moves it traces out an object. This object is the phase space set.

Time Series

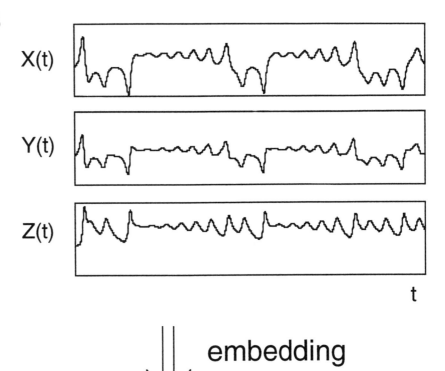

X(t)

Y(t)

Z(t)

t

embedding

Phase Space

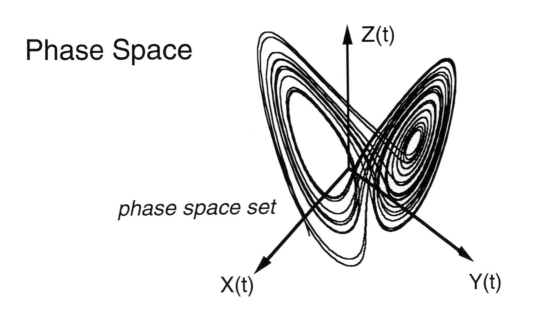

phase space set

131

Attractors

The coordinates of each point in the phase space correspond to one combination of possible values of the properties that could be measured for a system.

As time goes by, the values of the properties measured for a chaotic system take on only certain combinations of values. These combination of values correspond to a region in the phase space. This region is called an **attractor**.

If we start the system with an unusual set of values of the measured properties that are different from one of these preferred combinations, the values rapidly change to one of these preferred combinations. In the phase space, the unusual starting values correspond to a point away from the attractor. Thus, in the phase space, as time goes by, a point off the attractor rapidly approaches the attractor. That is why it is called an attractor.

The motion in the phase space before the system reaches the attractor is called the **transient** behavior. If we analyze the system after this transient behavior has ended, then the phase space set that we find corresponds to the attractor.

When the fractal dimension of an attractor is not an integer, then the attractor is said to be "**strange**."

Attractors in Phase Space

Logistic Equation

$X(n+1) = 3.95\, X(n)\, [1-X(n)]$

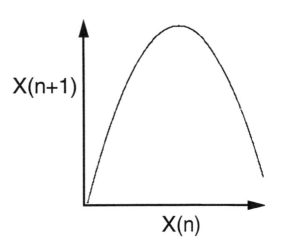

Lorenz Equations

$$\frac{dX}{dt} = 10\,(Y - X)$$

$$\frac{dY}{dt} = -XZ + 28X - Y$$

$$\frac{dZ}{dt} = XY - (8/3)Z$$

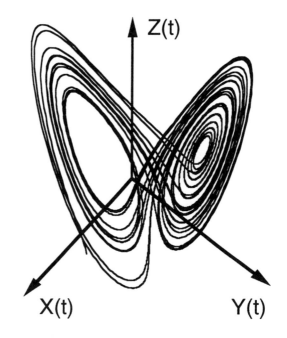

The Fractal Dimension of the Attractor Tells Us the Number of Independent Variables Needed to Generate the Time Series

The fractal dimension of the attractor in the phase space provides very useful information about the nature of the process that generated the sequence of values measured in time.

In a deterministic system, the present values of the measured properties are related to their previous values. The dimension of the attractor tells us the number of independent variables in this relationship. **The number of independent variables is the smallest integer that is greater than or equal to the fractal dimension of the attractor.**

For example, the fractal dimension of the attractor for the logistic system is slightly less than 1. Thus 1 equation with 1 independent variable can generate the sequence of values in time of the logistic system.

For example, the fractal dimension of the attractor for the Lorenz system is equal to 2.03. Thus a set of 3 equations with 3 independent variables can generate the sequence of values in time of the Lorenz system.

The number of independent variables = smallest integer ≥ the fractal dimension of the attractor

Logistic Equation

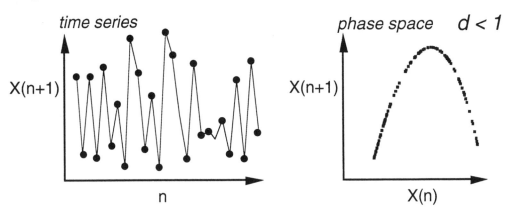

*d < 1, therefore, the equation of the time series that produced this attractor depends on **1** independent variable.*

Lorenz Equations

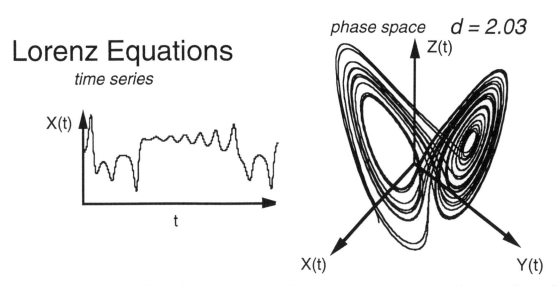

*d = 2.03, therefore, the equations of the time series that produced this attractor depends on **3** independent variables.*

Random is High Dimensional;
Chaos is Low Dimensional

The dimension of the phase space set reveals whether a random-looking set of data was generated by a random or by a deterministic mechanism. The dimension determines the number of equations and independent variables needed to generate the data. When the dimension is infinite, the data were generated by a random mechanism. When the dimension is small, the data were generated by a deterministic mechanism.

(There is no rigorous limit on what value of the dimension corresponds to "small." In practice, it is often difficult to tell whether the dimension of the phase space set is finite or infinite. Depending on the amount of data, the largest finite values that can be detected are typically between 3 and 6. This suggests that "small" means a value less than about 6.)

1. Random: Infinite Dimensional Phase Space Set

When the dimension of the phase space set is very large, then the sequence of values measured in time was generated by a random mechanism based on chance. That is, the number of equations and variables is so large that there are no rules that can be used to predict the future values from the past values.

What we mean by random is that so many different things are happening at once that it is not possible to understand how these mechanisms work, and thus we say that the future values are determined by chance.

2. Deterministic Chaos: Low Dimensional Phase Space Set

When the dimension of the phase space set is small, then the sequence of values measured in time was generated by a deterministic mechanism based on a small number of independent variables.

What we mean by determinism is that a small number of equations and their variables describe how the values in the past can be used to compute the values in the future.

136

Data 1

Data 2

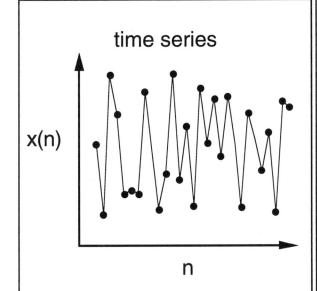

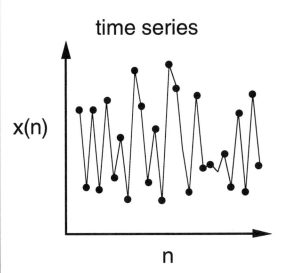

time series

time series

x(n)

x(n)

n

n

phase space

d → ∞

phase space

d = 1

Since d → ∞,
the time series
was produced by
a random
mechanism.

Since d = 1
the time series
was produced by
a deterministic
mechanism.

Constructing the Phase Space Set
from a Sequence of Data in Time

1. Experimental Measurement of All the Variables

The values of each of the relevant properties X, Y, Z, . . . are measured simultaneously over time. One point in the phase space is plotted for each set of measurements made at one time. The coordinates of each point are the values of each property X(t), Y(t), Z(t), . . . measured at time t.

2. Experimental Measurement of One Variable and Takens' Theorem

The values of one property X are measured over time. **Takens' theorem** says that the entire phase space set can be constructed from one independent variable. This remarkable procedure works because the variables are linked together by the relationship that produces the attractor.

One point in a phase space of dimension N is plotted for each measurement at time t. The coordinates of each point are X(t), X(t+Δt), X(t+2Δt), . . ., X(t+(N-1)Δt). The value Δt is called the **lag**. These lagged coordinates generate an N-dimensional phase space set.

(A "real" phase space would consist of the coordinates X(t), dX(t)/dt, d^2X(t)/dt^2, ... The coordinates X(t), X(t+Δt), X(t+2Δt), . . are a linear combination of the differences that approximate these derivatives. Thus Δt must be chosen so that the differences such as [X(t+Δt)-X(t)]/Δt are a good approximation of the derivatives. The fractal dimension of the phase space set is invariant under a linear transformation of the coordinates. Thus the fractal dimension of the phase space set computed in this way is equal to the fractal dimension of the "real" phase space set.)

The fractal dimension of the phase space set cannot be larger than the embedding dimension of the space that we put it into. Your 3-dimensional body casts only a 2-dimensional shadow on a 2-dimensional surface. Thus the embedding must be repeated for increasing N and the "real" dimension of the phase space set is the limiting value found as the embedding dimension N is increased.

Phase Space

Constructed by direct measurement:

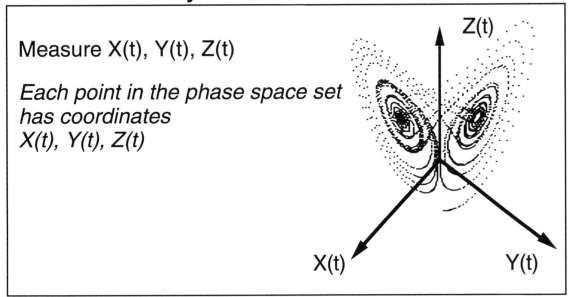

Measure X(t), Y(t), Z(t)

*Each point in the phase space set
has coordinates
X(t), Y(t), Z(t)*

Constructed from one variable:

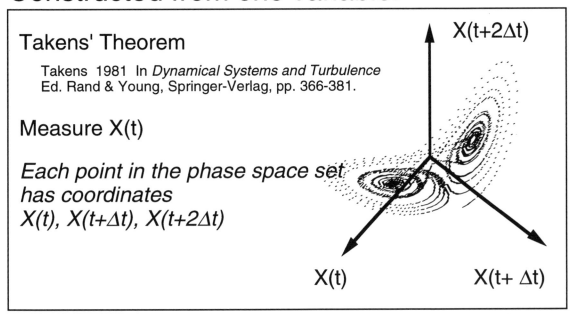

Takens' Theorem

Takens 1981 In *Dynamical Systems and Turbulence*
Ed. Rand & Young, Springer-Verlag, pp. 366-381.

Measure X(t)

*Each point in the phase space set
has coordinates
X(t), X(t+Δt), X(t+2Δt)*

Example of Phase Space Sets Constructed from Direct Measurement:
Motion of the Surface of Hair Cells in the Ear

The hair cells in the inner ear convert sound into electrical signals. Each cell has a hair that is deflected by sound. The motion of this hair alters the shape of the cell and the electrical properties of the cell membrane. These changes produce an electrical signal in response to the sound.

Teich et al. used light reflected off the surface of an individual hair cell to measure the *position* and *velocity* of the cell surface. Their measurement technique, called heterodyne optical interferometry, makes it possible to measure the position of the cell surface to an accuracy of 10^{-13} meter, which is about $1/1000$ the diameter of an atom. They used the change in frequency of the reflected light, called the Doppler shift, to measure the velocity of the surface of the cell.

They constructed phase space sets **directly** from the **measured position** and **velocity** of the surface of the hair cells. They analyzed the shape of these phase space sets and their fractal dimension. Each hair cell responds to sounds within a narrow range of frequencies around its natural frequency. They found that both the shape of the phase space set and its fractal dimension depended on the natural frequency of the cell and the frequency of the sound they used to stimulate the cell.

The hair can deflect more easily toward one direction than the opposite direction. They compared their experimental results to mathematical models where the mass or stiffness of the hair is greater in one direction than in the opposite direction. These models have provided information on the relationship between the motion of the hair and the electrical signals generated by these cells.

Position and Velocity of the Surface of a Hair Cell in the Inner Ear

Teich et al. 1989 Acta Otolaryngol (Stockh), Suppl. 467:265-279.

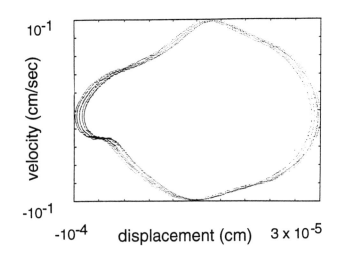

stimulus = 171 Hz

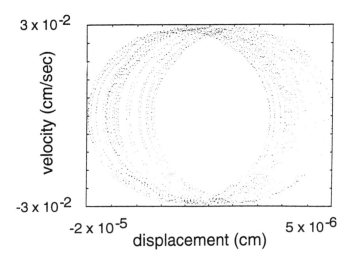

stimulus = 610 Hz

Example of Phase Space Sets Constructed from the Measurement of One Variable

Takens' theorem can be used to construct the phase space set from the measurement of the series of values of one variable x(t) in time t. The points in the N-dimensional phase space set have coordinates x(t), x(t+Δt), x(t+2Δt), . . . , x(t+(N-1)Δt). A series of such N-dimensional phase spaces are constructed with increasing N. If the dimension of the phase space set increases with increasing N, then the series of values x(t) was generated by a random mechanism. If the dimension of the phase space set reaches a constant value with increasing N, then the series of values x(t) was generated by a deterministic mechanism.

For example, this procedure was used to analyze the time series x(n) in Data Set #1 that was generated by the random mechanism of choosing the value of x(n) at random and Data Set #2 that was generated by the deterministic mechanism that x(n+1) = 3.95 x(n) [1 - x(n)]. The lag Δt was set equal to the time between consecutive points. The 2-dimensional phase space set was constructed from points with coordinates X=x(n) and Y=x(n+1). The 3-dimensional phase space set was constructed from points with coordinates X=x(n), Y=x(n+1), Z=x(n+2). And so on.

1. Data Set #1: Random

The **fractal dimension** of the **phase space set** increases as the embedding dimension increases. That is, the fractal dimension is **infinite**.

Thus this time series was generated by a random mechanism. That is, it was generated by a mechanism with an **infinite** set of independent variables. This is what we mean by random, that there is a very large number of different things happening at once.

2. Data Set #2: Deterministic Chaos

The **fractal dimension** of the **phase space set** reaches a limiting value slightly less than **1** as the embedding dimension increases.

Thus, this time series was generated by a deterministic rule that can be described by **1** equation with **1** independent variable.

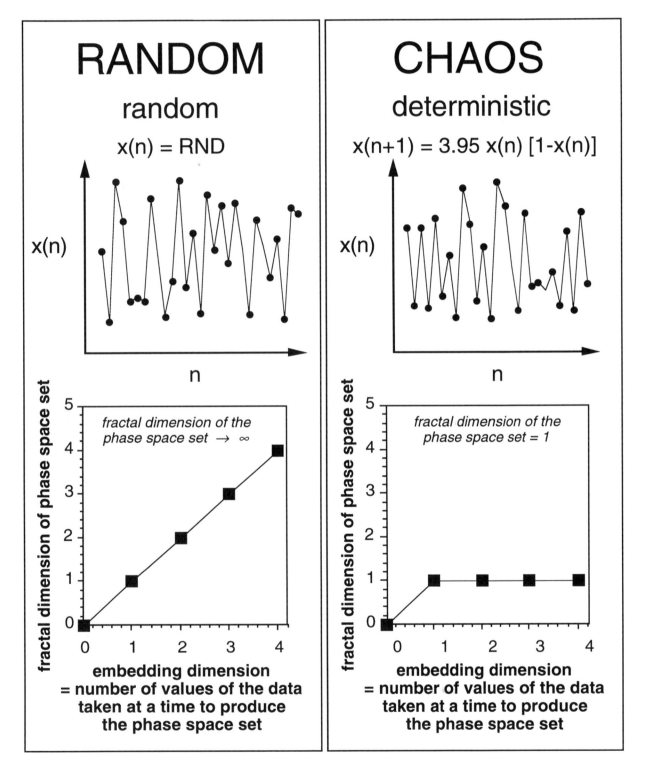

The Beating of Heart Cells

Glass et al. maintained chick heart cells alive in culture. They measured the electrical voltages produced by these cells. Since the voltage changes when the cells beat, they could determine the timing between beats. They also electrically stimulated the cells to beat.

Even when these cells are not stimulated, they beat spontaneously.

When they electrically stimulated the cells at a fast rate, then the cells could not recycle fast enough to beat in response to each stimulation. However, the cells did beat in response to every other stimulation.

When they electrically stimulated the cells at a slower rate, then the cells beat in response to each stimulation.

When they electrically stimulated the cells at an even slower rate, then there was enough time for the cells to beat between the stimulations. In the time during which 2 stimulations were given, the cells beat 3 times.

The number of stimulations was coupled to the number of beats in the integer ratios of 2:1, 1:1, and 2:3. Such an integer coupling that depends on the stimulus frequency is a hallmark of a nonlinear response. Linear systems may respond with different phase lags when stimulated at different frequencies, but the type of response, here the number of beats, does not vary with frequency.

Chick Heart Cells

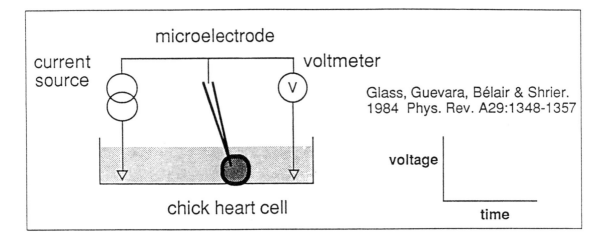

microelectrode

current source   voltmeter

Glass, Guevara, Bélair & Shrier.
1984 Phys. Rev. A29:1348-1357

voltage

time

chick heart cell

Spontaneous Beating, No External Stimulation

Periodically Stimulated

2 stimulations – 1 beat

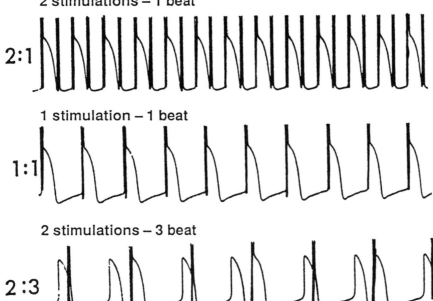

2:1

1 stimulation – 1 beat

1:1

2 stimulations – 3 beat

2:3

Example of Using the Phase Space Set to Differentiate Random and Deterministic Mechanisms: The Beating of Heart Cells

When Glass et al. electrically stimulated chick heart cells in culture at a certain rate, the timing between the beats of the heart cells seemed to occur at random. Was the timing between the beats generated by a random mechanism or by a deterministic mechanism?

They measured the time ϕ between each stimulation and the subsequent beat number n. They constructed the phase space set from this time sequence $\phi(n)$. Each point in the phase space set had coordinates $X=\phi(n)$ and $Y=\phi(n+1)$.

If the timing between beats was generated by a random process based on chance, then the phase space set would fill this 2-dimensional phase space. **The phase space set did not fill the 2-dimensional phase space.** The phase space set was approximately 1-dimensional. This means that the timing between the beats can be described by a **deterministic** relationship. They derived the mathematical form of this relationship. It depended on the times of the previous beat and the previous stimulation. They showed that this relationship arises from the properties of the refractory period when the heart cell is recycling after a beat.

These experiments probe the biochemical and electrical processes that occur when a heart cell recycles after a beat. They are also a model of the whole heart. The stimulations correspond to the pulses from the sinoatrial node that initiates the heartbeat, and the cells correspond to the ventricle, the main pump of the heart. Thus the equations that describe this experiment may also shed light on the causes of irregular heart rhythms. Present methods control such problems by using large electrical currents to force the heart out of its irregularities. If such irregular heart rhythms are deterministic, they might be controlled by much smaller electrical currents applied at appropriate times computed from the deterministic relationship.

The Pattern of Beating of Chick Heart Cells

Glass, Guevara, Bélair & Shrier. 1984. Phys. Rev. A29:1348-1357

periodic stimulation – chaotic response

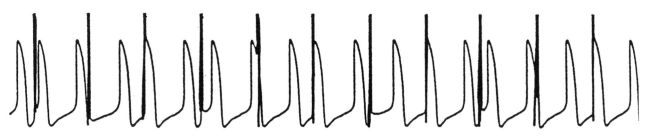

φ = phase of the beat with respect to the stimulus

phase vs. previous phase

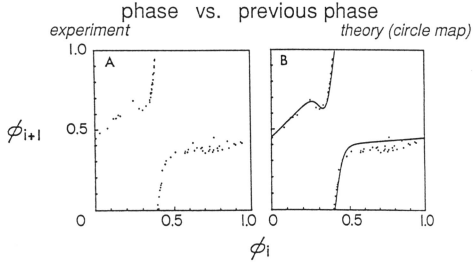

Since the phase space set is 1-dimensional,
the timing between the beats of these cells
can be described by a <u>deterministic</u> relationship.

Overview of the Phase Space Analysis

Measure a sequence of values in time.

Turn the time series into an object in space. This procedure is called an **embedding**. The space is called the **phase space**. The object is called the **phase space set**.

Determine the **fractal dimension** of the phase space set.

If the fractal dimension of the phase space set is **large**, then the mechanism that generated the data is a **random** mechanism that is based on **chance**.

If the fractal dimension of the phase space set is **small**, then the mechanism that generated the data is **deterministic**.

If the mechanism that generated the data is deterministic, but the time series mimics that produced by a random process, then the system is said to be "**chaotic**."

Procedure

1. **Time Series**

   e.g. voltage as a function of time

2. **Turn the Time Series into a Geometric Object**

   This is called <u>embedding</u>.

3. **Determine the Topological Properties of this Object**

   Especially, the <u>fractal dimension</u>.

4. **High Fractal Dimension**
   = **Random** = chance
   Low Fractal Dimension
   = **Chaos** = deterministic

The Fractal Dimension
Is <u>Not</u> Equal to
the Fractal Dimension

Unfortunately the **same** words, "the fractal dimension of the time series" have been used to refer to two different concepts.

Each of these "fractal dimensions" is a **different thing** and they **are not equal to each other**.

1. Fractal Dimension of the Time Series Itself

This fractal dimension characterizes the self-similarity of the time series itself.

The value of this fractal dimension describes how the small variations in the values measured over brief time intervals are related to the large variations measured over long time intervals.

For example, it can be computed by using the box counting algorithm to analyze the plot of the time series values $x(t)$ plotted versus the time t.

2. Fractal Dimension of the Phase Space Set Constructed from the Time Series

This fractal dimension characterizes the properties of the phase space set, not the original time series.

The value of this fractal dimension tells us the number of independent variables needed to generate the time series from which the phase space was constructed.

For example, it can be computed by using the box counting algorithm to analyze the phase space set constructed from the points with the coordinates $x(t)$, $x(t+\Delta t)$, $x(t+2\Delta t)$, . . . , $x(t+(N-1)\Delta t)$, where $x(t)$ are the time series values, Δt is the lag, and N is the dimension of the phase space.

The Fractal Dimension
is **NOT** equal to
The Fractal Dimension

Fractal Dimension:

How many new pieces of the Time Series are found when viewed at finer time resolution.

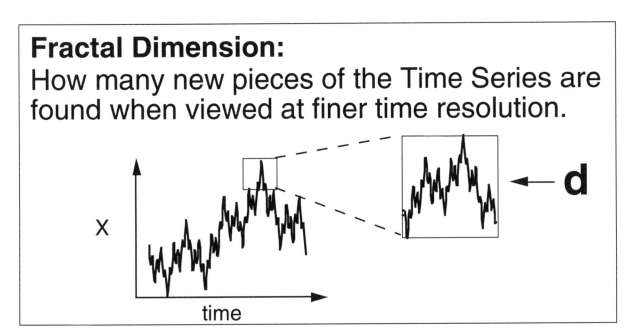

Fractal Dimension:

The Dimension of the Attractor in Phase Space is related to the Number of Independent Variables.

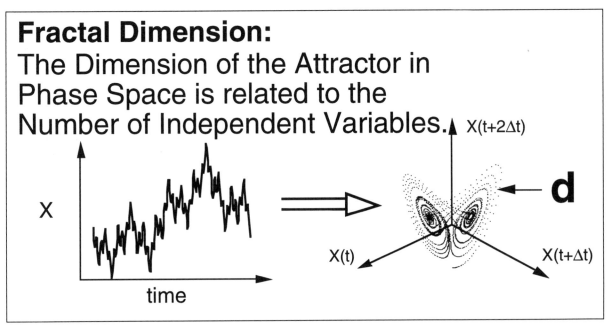

Biological Implications of the Phase Space Analysis

1. Chance or Necessity?

We are used to thinking that the variability found in biological systems is due to mechanisms based on chance that reflect random processes.

We now know that simple, deterministic mechanisms can also generate output that is so complex that it mimics the behavior generated by random processes.

Phase space analysis now makes it **possible to determine** if the data measured were generated by a mechanism based on a **random** process or by a mechanism based on a **deterministic** process.

When the fractal dimension of the phase space set is very large, then the data were generated by a mechanism based on a random process. When the fractal dimension of the phase space set is small, then the data were generated by a mechanism based on a deterministic process.

2. Understanding

If the variability found in biological systems is due to chance, there is little hope of ever fully understanding how these systems work.

However, if the variability in the data is due to the fact that a system is chaotic and therefore deterministic, then we may be able to fully **understand**, and even **control** it.

Mechanism That Generated the Data

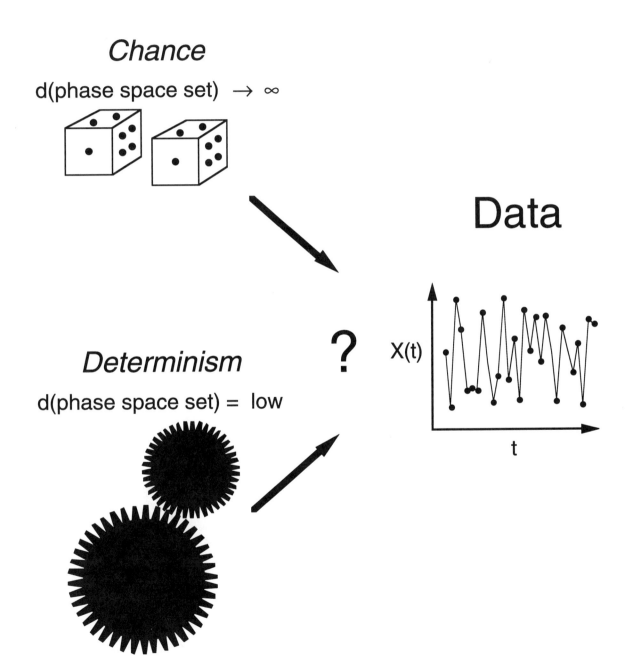

Chance

d(phase space set) → ∞

Determinism

d(phase space set) = low

?

Data

X(t)

t

CHAOS

Sensitivity to Initial Conditions

155

Lorenz System: Physical Description

We use the Lorenz system to introduce the concept of **sensitivity to initial conditions**.

Lorenz studied the properties of a simplified model of the motion of the air in the atmosphere. The air is heated from below and cooled from the top. Hot air rises and cold air falls. The air rises and falls along the opposite edges of long cylinders. It is as if the cylinders of air are rotating. At the beginning one cylinder is rotating clockwise. As it turns, it brings hot air up from the bottom into the cold air at the top. It also brings cold air down from the top into the hot air at the bottom. Thus it mixes the air, reducing the temperature difference in the air. This temperature difference is the force driving the motion of the air. Hence, as the cylinder turns, it reduces the force driving its own motion. Thus the motion of the cylinder slows down and stops. However, without the motion of the air, the heat imposed from the bottom and the cold imposed from the top soon builds up the temperature difference in the air once again. Thus the cylinder starts to rotate again. Only now, it rotates in the opposite direction, counterclockwise. It rotates counterclockwise for a while. Again, as it rotates it reduces the force driving its own motion. It stops, the temperature difference builds up, and now it switches directions again, and rotates clockwise.

In summary, the Lorenz system consists of the motion of air along cylinders that rotate first clockwise, then counterclockwise, then clockwise, and keep switching from one direction to the other.

This system can be approximated by 3 equations and 3 independent variables. We will consider only the variable X. The value of X is the amount of angular velocity. That is, when X>0, a cylinder of air rotates clockwise; and when X<0, a cylinder of air rotates counterclockwise.

A phase space can be constructed from the variables X(t), Y(t), and Z(t). At time t, the system is represented by a point in the phase space with coordinates X(t), Y(t), and Z(t).

Lorenz

1963 J. Atmos. Sci. 20:13-141

Model

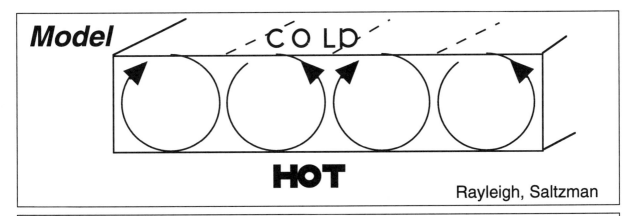

COLD

HOT

Rayleigh, Saltzman

Equations

$$\frac{dX}{dt} = 10 (Y - X) \qquad \frac{dY}{dt} = -XZ + 28X - Y \qquad \frac{dZ}{dt} = XY - (8/3)Z$$

X = speed of the convective circulation
      X > 0   clockwise,   X < 0   counterclockwise

Y = temperature difference between rising and falling fluid

Z = bottom to top temperature minus the linear gradient

Phase Space

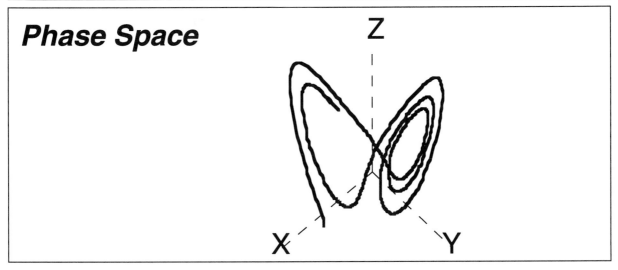

Lorenz System: Phase Space Set

The phase space set of the Lorenz system does not fill the 3-dimensional phase space. The region occupied by the phase space set is the **attractor**. The fractal dimension of the attractor is approximately 2.03. Since this attractor has a fractal dimension that is not equal to an integer, this attractor is called a **strange attractor**.

The low fractal dimension of the attractor reveals the fact that the Lorenz system is deterministic. The number 3 is the smallest integer greater than or equal to 2.03. This reveals that the Lorenz system consists of 3 equations with 3 independent variables.

The phase space set of the Lorenz system looks like a butterfly.

The right wing of the butterfly is in the part of the phase space where $X>0$. Thus all the points on this wing correspond to the cylinder rotating clockwise. The left wing of the butterfly is in the part of the phase space where $X<0$. Thus all the points on this wing correspond to the cylinder rotating counterclockwise.

Each set of values of $X(t)$, $Y(t)$, and $Z(t)$ measured for the physical system corresponds to one point in the phase space. As the physical system evolves in time, the point representing it moves through the phase space along the butterfly. For example, if the point starts on the right wing, $X>0$, the cylinder is rotating clockwise. The point loops around the right wing a number of times until it switches to the left wing. On the left wing, $X<0$, and the cylinder is rotating counterclockwise. The point loops around the left wing a number of times until it switches back to the right wing. Back on the right wing, $X>0$, and the cylinder is rotating clockwise again.

The figure does not do justice to the fine structure of the attractor. The more the attractor is magnified, the more lines are revealed that trace out the motion of the point in phase space that represents the evolution of the state of the system in time. The attractor is fractal.

Lorenz Attractor

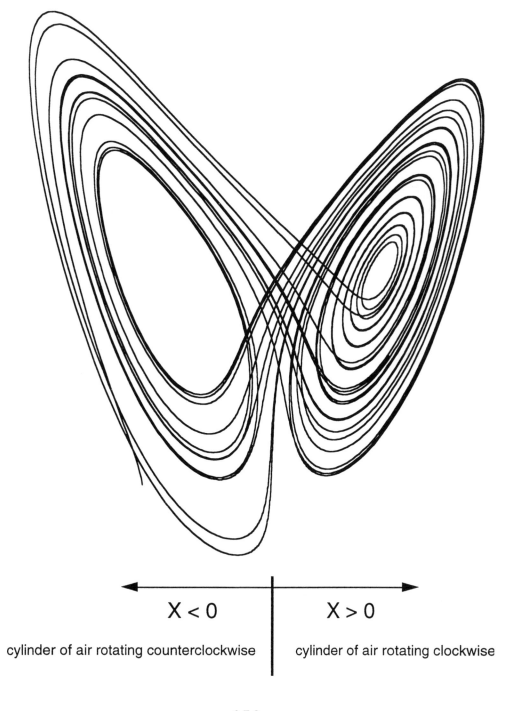

X < 0   X > 0

cylinder of air rotating counterclockwise   cylinder of air rotating clockwise

Lorenz System: Sensitivity to Initial Conditions

Given a set of starting values for X, Y, and Z, we use the Lorenz equations to compute the values of the variables X(t), Y(t), and Z(t) at different times t. These starting values are called the **initial conditions**.   We first run the computation with one set of starting values and then rerun it a second time with the starting value of X changed ever so slightly.

1. Initial Condition: X=1

First we compute the values of X(t) from the starting values X=1, Y=1, and Z=1.

The starting value of X=1 is greater than zero. Thus the cylinder of air starts off rotating clockwise. After a short time X becomes less than zero and the cylinder is rotating counterclockwise. Then X becomes greater than zero and the cylinder is rotating clockwise again. As time goes by, X becomes less than or greater than zero, and so the rotation of the cylinder switches between counterclockwise and clockwise.

2. Initial Condition: X=1.0001

Then we compute the values of   X(t) from the starting values X=1.0001, Y=1, and Z=1.

That is, we now rerun the same computation, just starting the initial value of X ever so slightly differently. At the beginning, the values of X(t) closely follow the values in the first run. This is not surprising, because the initial values of X were so similar.

However, after a while, at the same elapsed time, when X>0 in the first run, now X<0 in the second run. *Even though the two runs were started with almost identical initial conditions, now the state of the system in the second run is completely different from the state of the system at the same time in the first run.* That is, the cylinder of air that is rotating counterclockwise in the second run was rotating clockwise at the same time in the first run.   This is called **sensitivity to initial conditions**.

The Lorenz system amplifies small differences in the initial conditions into large differences in the values of the variables later on. The time it takes for these large differences to appear is equal to $(1/\lambda)$, where λ is called the **Liapunov exponent**.

Sensitivity to Initial Conditions
Lorenz Equations

Initial Condition:

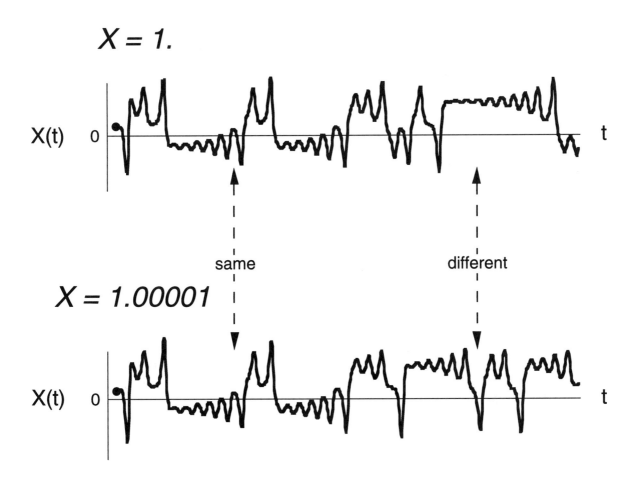

$X = 1.$

$X = 1.00001$

$$|X_{top}(t) - X_{bottom}(t)| \propto e^{\lambda t}$$

λ = Liapunov Exponent

Chaos: Deterministic but Not Predictable

1. Non-Chaotic: Deterministic and Predictable

For a deterministic system that is *not* chaotic, the value of a variable X at a given time can be used to calculate accurately the value of X at all times in the future.

2. Chaotic: Deterministic and Not Predictable

For a deterministic system that is chaotic, the value of a variable X at a given time can be used to calculate the value of X at a *brief* time in the future.

However, as we calculate the value of X at longer times into the future, the **sensitivity to initial conditions** means that the calculation becomes less accurate.

For example, if the starting value of X was known to 5 significant digits, as we calculate the value of X at longer times into the future, the number of accurate digits in the result will become 4, then 3, then 2, then 1, then zero. That is, the values of X at long times into the future cannot be predicted.

If we could know the values of the initial conditions to infinite accuracy, we could predict the future values of X. However, we can only measure the initial conditions to finite accuracy, for example, 5 significant digits. The sensitivity to initial conditions amplifies this initial inaccuracy so that it is not possible to predict the values of X far into the future.

A chaotic system is deterministic but not predictable in the long run. It is deterministic because the values of the variables at the next instant in time are determined by their present values. However, it is not predictable in the long run, because the sensitivity to initial conditions amplifies the inaccuracy that is always present in the initial conditions so that we cannot accurately predict the values of the variables far into the future.

Deterministic, Non-Chaotic

X(n+1) = f {X(n)}

Accuracy of values computed for X(n):

1.736 → 2.345 → 3.254 → 5.455 → 4.876 → 4.234 → 3.212

Deterministic, Chaotic

X(n+1) = f {X(n)}

Accuracy of values computed for X(n):

3.4552 → 3.45? → 3.4?? → 3.??? → ? → ? → ?

The Clockwork Universe and
The Chaotic Universe

It used to be thought that the Universe was like a great clock. If we could learn the equations that described the mechanisms of the great clock and if we knew the values of its variables at the present, then we could predict all the future values of the variables.

However, if the mechanism of the Universe is chaotic, then we cannot predict all the future values of the variables. Even if we could learn the equations that described the mechanisms of the great clock and even if we knew the values of its variables at the present, then we could still not predict the future values of the variables. The uncertainty in this Universe does not arise from chance. It arises from the sensitivity to initial conditions.

These two contrasting views were eloquently stated by Laplace in 1776 and Poincaré in 1890.

1. Laplace: Clockwork Universe

"The present state of the system of nature is evidently a consequence of what it was in the preceding moment, and if we conceive of an intelligence which at a given instant comprehends all the relations of the entities of this universe, it could state the respective positions, motions, and general effects of all these entities at any times in the past or future."

2. Poincaré: Chaotic Universe

"A very small cause which escapes our notice determines a considerable effect which that we cannot fail to see, and then we say that the effect is due to chance. If we knew exactly the laws of nature and the situation of the universe at the initial moment, we could predict exactly the situation of that same universe at a succeeding moment. But even if it were the case that the natural laws had no longer any secret for us, we could still only know the initial situation approximately. If that enables us to predict the succeeding situation with the same approximation, that is all we require, and we should say that the phenomenon had been predicted, that it is governed by laws. But it is not always so; it may happen that small differences in the initial conditions produce very great ones in the final phenomenon. A small error in the former will produce an enormous error in the latter. Prediction becomes impossible, and we have the fortuitous phenomenon."

164

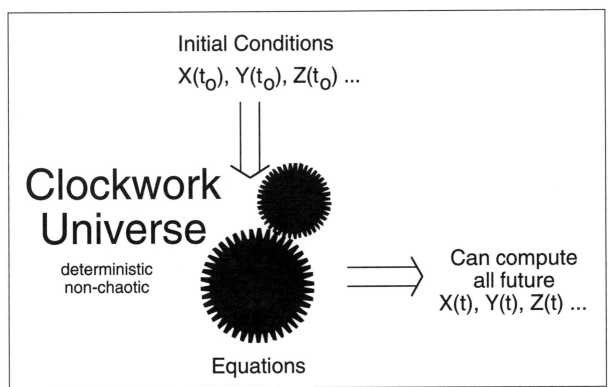

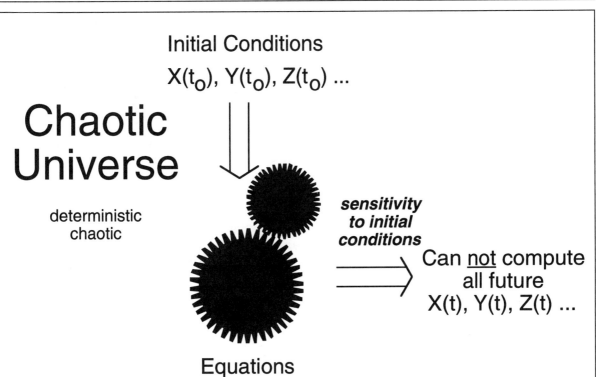

Lorenz System: A Strange Attractor

1. Points away from the Attractor Are Drawn toward It

The Lorenz attractor does not fill the 3-dimensional phase space. Only certain combinations of the values X, Y, and Z are present. We can push the system away from this set of combinations. For example, we can reach in and put all the hot air next to the cold plate on top and all the cold air next to the hot plate on the bottom. The system will then rapidly escape from this unnatural state back towards a more natural state. In the phase space, this unnatural state is represented by a **point off the attractor**. As the system evolves toward a more natural state, this point **rapidly approaches the attractor**.

2. Points on the Attractor Pull away from Each Other

Sensitivity to initial conditions can be seen in the motion of **points on the attractor**.

We start the system with a combination of values of X, Y, and Z that are part of the attractor. If we then rerun the system a second time with ever so slightly different values of these initial conditions, then the point representing the original run and the point representing the rerun will **rapidly separate from each other**. This is a reflection of the sensitivity to initial conditions. The two points separate exponentially fast from each other. For example, we could start both runs on the right wing where X>0, and the cylinder of air is rotating clockwise. After the same elapsed time, the point of the first run has switched to the left wing, where X<0, and the cylinder of air is rotating counterclockwise while the point of the second run is still on the right wing, where X>0, and the cylinder of air is rotating clockwise.

Although they separate from each other, both points remain on the attractor. The attractor is of a limited size. Thus they cannot separate from each other indefinitely. What happens is that after a while they are folded back onto each other. They separate and they are folded back onto each other, again and again. Thus the finer we look, the more of these trails we find. The evolution of the system in time that is represented by the trail of the point in the phase space forms the fractal structure of the attractor.

Lorenz Strange Attractor

Trajectories from outside: Trajectories on the attractor:
pulled <u>TOWARDS</u> it   pushed <u>APART</u> from each other

why its called an attractor   *sensitivity to initial conditions*

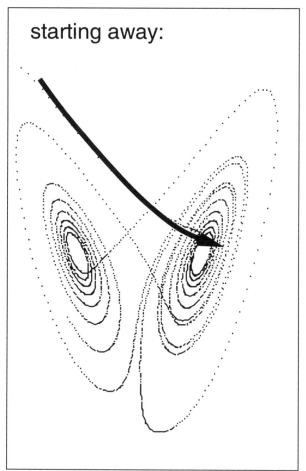

starting away:

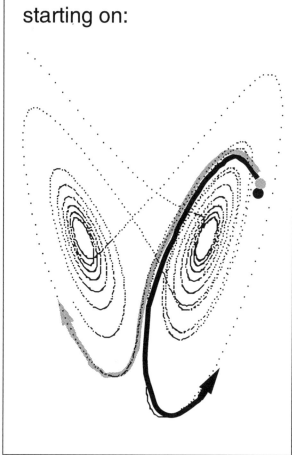

starting on:

"Strange" and "Chaotic"

1. "Strange" Means that the Attractor Is Fractal

The formal definition of "strange" is that the fractal dimension of the attractor is not an integer.

2. "Chaotic" Means Sensitivity to Initial Conditions

The formal definition of "chaotic" is that if the system is rerun with almost the same starting conditions, the values of the variables measured at the same time of the two runs separate from each other exponentially fast as a function of time.

All 4 combinations are possible:
    The Lorenz system is strange and chaotic.
    There are systems that are not strange and not chaotic.
    There are systems that are strange and not chaotic.
    There are systems that are not strange and chaotic.

"Strange"

attractor is fractal

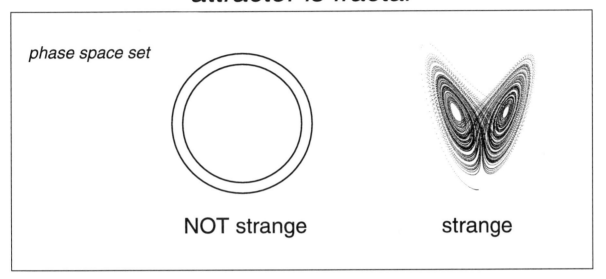

phase space set

NOT strange   strange

"Chaotic"

sensitivity to initial conditions

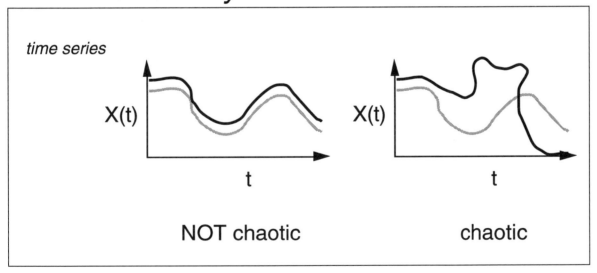

time series

X(t)

t

NOT chaotic   chaotic

The Shadowing Theorem

The coordinates of a point in the phase space set of the Lorenz system at time t are equal to the values of the variables $X(t)$, $Y(t)$, and $Z(t)$. As the system evolves in time, the values of these variables change. Thus the point moves in the phase space. The line traced out by the moving point is called a **trajectory**. These trajectories form the attractor. The values of the variables can be found by numerical integration of the Lorenz equations. That is, the values of the variables at the next point in time are computed from their values at the previous point in time.

There are always some errors in the values of variables computed by numerical integration. Sensitivity to initial conditions means that the Lorenz system will amplify these small errors into large differences in the values of the variables. In fact, sensitivity to initial conditions means that after a while, the values of these variables are unpredictable. *So how can we compute the shape of the attractor?*

The answer is subtle and beautiful. It is called the **shadowing theorem**. This theorem says that the errors introduced by the sensitivity to initial conditions mean that we did not accurately compute the trajectory that began with the initial conditions that we used. However, the values of the variables that we did compute, errors and all, are a good approximation of another "real" trajectory on the attractor. More formally, there is a "real" trajectory that "shadows" (is close to) the one that we computed. This "real" trajectory is one that has a set of initial conditions different from the ones that we used.

The reason for this wonderful result is that if the errors push the values of the variables off the attractor, they will be rapidly drawn back to their values on the attractor. If the errors push the values of the variables to another set of values on the attractor, then we just pick up one of the infinite set of other trajectories on the attractor.

Shadowing Theorem

If the errors at each integration step are small, there is an EXACT trajectory which lies within a small distance of the errorfull trajectory that we calculated.

There is an INFINITE number of trajectories on the attractor.
When we go off the attractor, we are sucked back down exponentially fast.
We're on an exact trajectory, just not the one we thought we were on.

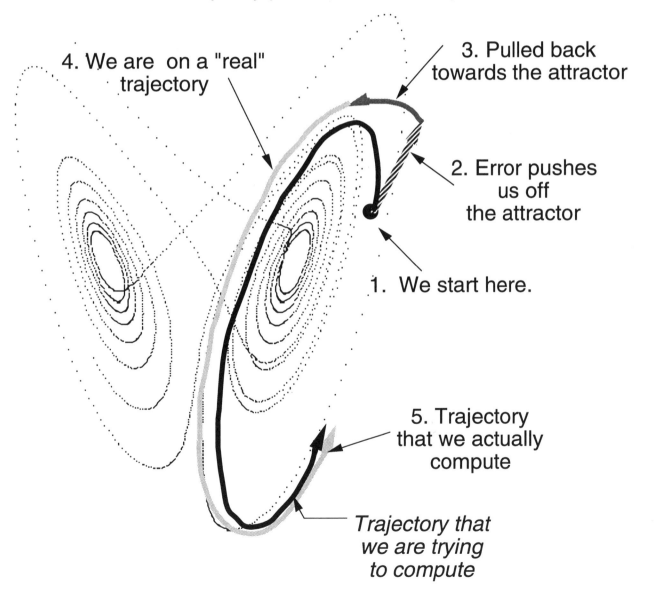

4. We are on a "real" trajectory

3. Pulled back towards the attractor

2. Error pushes us off the attractor

1. We start here.

5. Trajectory that we actually compute

Trajectory that we are trying to compute

Biological Implications of Sensitivity to Initial Conditions

Suppose that we take a dish of cells out from our incubator on Tuesday, add 10 μl of some drug into the dish, and find that it produces a certain effect in the cells. We are used to thinking that if we take out another dish of cells from our incubator on Wednesday, add 10 μl of the same drug into the dish, that we should see approximately the same effect in the cells.

This type of thinking is based on our experience with *linear* systems. This behavior is not necessarily true for *nonlinear systems*. It is *not* true for *chaotic systems.*

Chaotic systems have sensitivity to initial conditions. This means that quite similar values for the starting conditions will produce quite different results.

The variability found when experiments on biological systems are repeated has previously been attributed to chance, or to the complexity of these systems, or to uncontrollable factors in the experiment.

However, those may not be the only possible reasons for the observed variability found in experiments on biological systems. Perhaps some biological systems are chaotic, and this **variability** is a manifestation of their **sensitivity to initial conditions**.

Sensitivity to initial conditions means that the conditions of an experiment can be quite <u>similar</u>, but that the results can be quite <u>different</u>.

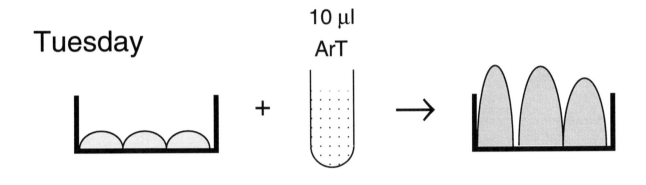

Tuesday 10 μl ArT

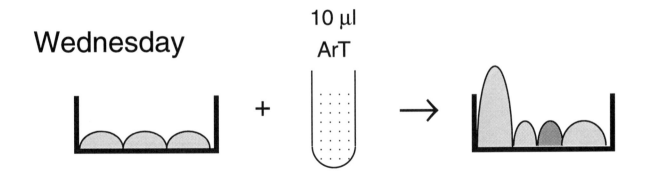

Wednesday 10 μl ArT

CHAOS

Bifurcations

175

Bifurcation: An Abrupt Shift in Behavior

The behavior of a system may depend on the values of certain properties of the system. These values are called **parameters**.

You might expect that when the value of a parameter changes by a small amount the behavior of the system also changes in a small way.

However, sometimes, when the value of a parameter changes by a small amount, there is a large change in the behavior of the system. This is called a **bifurcation**.

$$x(n+1) = A\,x(n)\,[1-x(n)]$$

A = 3.22

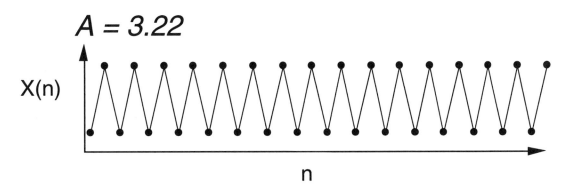

X(n)

n

A = 3.42

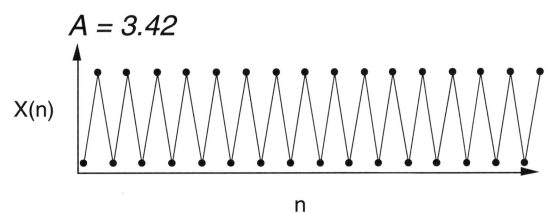

X(n)

n

– – – – – – – – – – – – – – *Bifurcation* – – – – – – – – – – – – –

A = 3.62

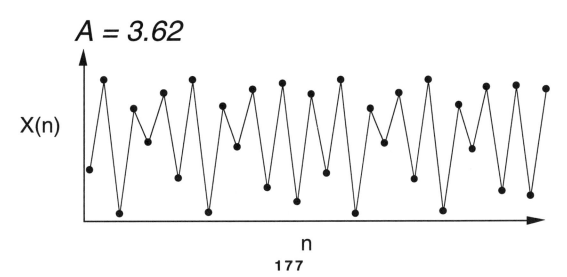

X(n)

n

Bifurcation Diagrams

One way to detect a bifurcation is to plot values of one of the variables of the system versus the value of a parameter. This plot is called a **bifurcation diagram**.

Each point on the x-axis corresponds to one value of the parameter. Vertically above this parameter is then plotted many values of the variable as the system evolves in time. The y-direction is therefore a smear of all the values of the variable that occur at that value of the parameter on the x-axis. This vertical smear forms a pattern. An abrupt change in the form of this pattern with a small change in the parameter plotted on the x-axis signals the existence of a bifurcation.

The bifurcation diagram of a chaotic system has fractal features. That is, small regions of the plot have the same structure as larger regions.

$$x(n+1) = A\ x(n)\ [1-x(n)]$$

Start with one value of A.
Start with x(1) = .5.
Use the equation to compute x(2) from x(1).
Use the equation to compute x(3) from x(2)
 and so on... up to x(300).
Ignore x(1) to x(50), these are the transient
 values off of the attractor.
Plot x(51) to x(300) on the Y-axis
 over the value of A on the X-axis.
Change the value of A,
and repeat the procedure again.

Sudden changes of the pattern indicate bifurcations (↑).

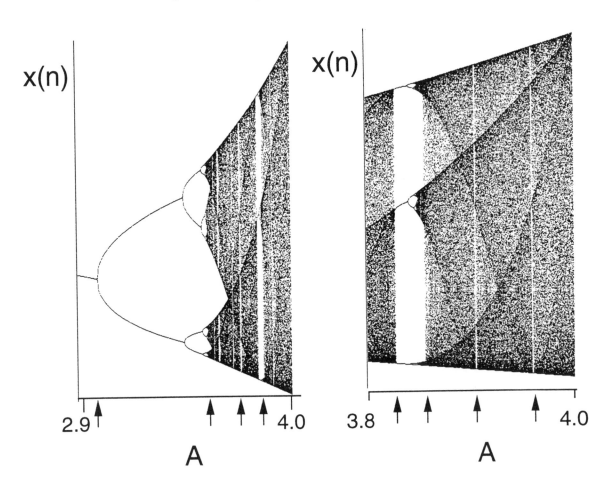

Example of Bifurcations:
Theory of Glycolysis

Energy is stored in sugar molecules. Biochemical reactions in the body transfer this energy from the sugar to a molecule called ATP, adenosine triphosphate. ATP is then used as an energy source to drive many other biochemical reactions in the body. One process that accomplishes this transfer of energy from sugar to ATP is called **glycolysis**.

There are a number of reactions in glycolysis. The overall speed of the reaction system is set by two steps that involve enzymes. Each enzyme speeds up one important reaction. The products produced by each of these reactions also affect the rate of the enzyme itself. Thus there is positive and negative feedback control in this reaction system.

Markus and Hess formulated the set of equations that describe the reactions in glycolysis. They used these equations to determine how the ATP concentration in time would depend on different types of sugar input.

For example, they studied what would happen if the flow of sugar was input into these biochemical reactions in a periodic way. The parameter they varied was the frequency at which the sugar was input into the system. They found that for some frequencies the ATP concentration fluctuated in a *periodic* way. For other frequencies, the ATP concentration fluctuated in a *chaotic* way. There was a sudden change in behavior from periodic to chaotic fluctuations as the frequency at which the sugar was input into the system changed by a small amount. This sudden change of behavior as a parameter is varied is called a **bifurcation**.

Glycolysis
The energy in glucose is transfered to ATP.
ATP is used as an energy source to drive biochemical reactions.

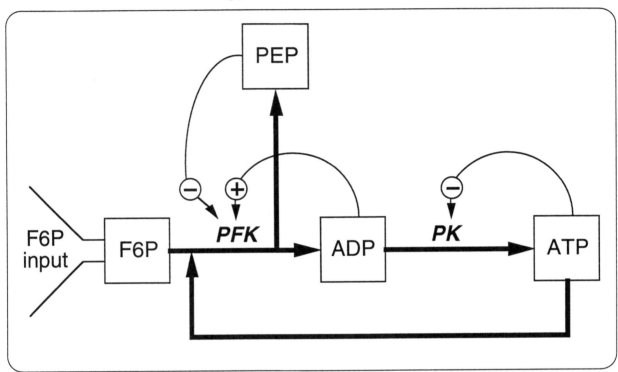

Theory   Markus and Hess 1985 Arch. Biol. Med. Exp. 18:261-271

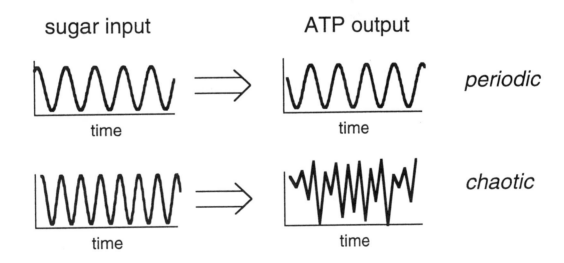

Example of Bifurcations:
Experiments of Glycolysis

The mathematical model developed by Markus and Hess predicted that there should be a bifurcation in the output ATP concentration, from periodic to chaotic behavior, as the frequency of the input flow of sugar was varied.

They tested their predictions using the contents from yeast cells to perform glycolysis in the solution in a glass container. They modulated the flow of sugar into the container at different frequencies. To determine how the ATP concentration varied in time, they measured the amount of blue light given off when a molecule that interacts with ATP is excited by ultraviolet light.

At low frequencies of sugar input, the experiments showed that the concentration of ATP varied periodically with time. However, at higher frequencies of sugar input, the concentration of ATP varied chaotically with time.

If you had found similar chaotic fluctuations in the data from your biological experiment, what do you think would happen if you tried to publish those results or get a grant to continue the work? People would look at those fluctuations and might say that you didn't know how to do those experiments properly. They might say that those fluctuations were due to experimental error. Perhaps your sensor was broken. Or they might say that the conditions were varying during the experiment. Perhaps the pH was changing in your solution. It is important to understand that the fluctuations that Markus and Hess found under certain conditions are *not* due to experimental error and they are *not* due to the conditions changing during the experiment. These variations are at the heart of what a chaotic system does all by itself.

Are all the variations we see in experiments due to chaos? Of course not. However, we now have the mathematical tools, such as the dimension of the phase space set and bifurcation diagrams, to analyze experimental data to determine if the variations are due to chance errors or deterministic chaos.

Glycolysis

Experiments   Hess and Markus 1987 Trends. Biochem. Sci. 12:45-48

cell-free extracts   ATP measured by fluoresence
from baker's yeast   glucose input   time

Periodic

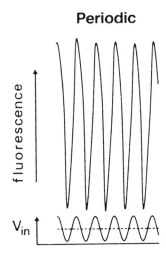

Chaotic

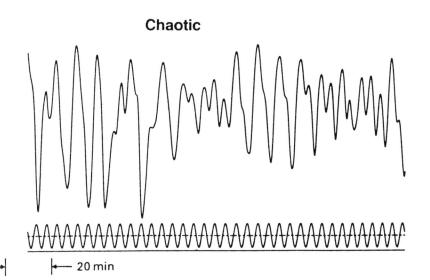

Example of Bifurcation Diagrams:
Theory and Experiments of Glycolysis

Markus et al. used a **bifurcation diagram** to compare their theoretical predictions with their experimental results. Each value on the horizontal axis corresponds to measurements at one value of the frequency of the input sugar flow. Along the vertical line above each point is plotted *the ATP concentration measured at the same time during each cycle of the input sugar flow.*

When there is only one point vertically over one frequency on the horizontal axis, it means that there was the same value of the ATP concentration at the same time in the input sugar cycle, and thus the variation of ATP concentration was **periodic**. When there are two points at a given frequency, it means that at one input sugar cycle the ATP concentration had one value, and that at the next input sugar cycle it had a different value. These same values then repeated on alternating input sugar flow cycles. When this was the case, then the ATP concentration took twice as long to repeat as the period of the input sugar flow cycle. This change in behavior is called a **period doubling bifurcation**. When the ATP concentration is spread along a vertical line, it means that there was a different value of the ATP concentration at each input sugar cycle, and thus the variation was **chaotic**.

The bifurcation plot predicted by the equations that model glycolysis is complex. At low frequencies, the period of the ATP concentration matches that of the input sugar flow. As the frequency of the input sugar flow increases, there are period doubling bifurcations, then a region of chaos, then a period 5 times as long as that of the input sugar flow cycle, then a second region of chaos, and finally a period 3 times as long as that of the input sugar cycle.

The experiments found the same sequence of bifurcations predicted by the equations. The locations of the bifurcations are somewhat shifted in frequency. Nonetheless, even the fine details of the order of the bifurcations in the experimental results match the theoretical predictions. This match tells us that this system is deterministic. The fluctuations present at certain frequencies are chaotic and are not generated by a random mechanism.

Glycolysis

Markus et al. 1985 Biophys. Chem. 22:95-105

Bifurcation Diagram

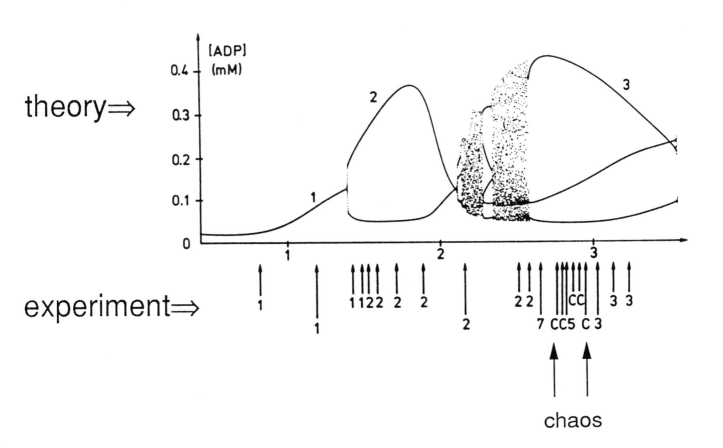

ADP measured at the same phase each time of the input sugar flow cycle

(ATP is related to ADP)

$$\# = \frac{\text{period of the ATP concentration}}{\text{period of the input sugar flow cycle}}$$

frequency of the input sugar flow cycle

Example of Bifurcations:
Motor and Sensory Phase Transitions

Some mechanical and electrical systems consist of many individual interacting parts. However, cooperation between the parts can lead to coherent behavior of the entire system, which Haken called *"synergetics."* This coherent behavior can change by a large amount when the value of a control parameter changes by a small amount. These changes have similar characteristics to *phase transitions,* such as the change of a substance from a liquid to a gas. For example, there is an increase in fluctuations as the control parameter approaches its value at the phase transition.

These phase transitions can be described by a potential energy function. The shape of this potential energy function depends on the control parameter. As the control parameter changes, the shape of the potential energy function changes, and thus the behavior of the system changes.

Kelso et al. showed that if you tap the index finger on your left hand in time with the tick of a metronome, you can also move the index finger on your right hand in the *opposite* direction. As the frequency of the metronome increases, the movement of your right finger suddenly shifts to motion in the *same* direction as that of your left finger. This change in behavior has the characteristics of a phase transition. They also found strong evidence of such bifurcations in the motor control of arms, hands, and legs.

Living systems have many components. It has been hard to understand how the activity of all these different nerves and muscles organize to produce coherent motion in our body. The finding of these phase transitions in the human body suggests that biological components interact in a way that self-organizes them into coherent function. Determining how a measured property depends on a control parameter that can be varied in an experiment gives us a way to test for this type of synergistic self-organization.

Phase Transitions

Haken 1983 *Synergetics: An Introduction* Springer-Verlag
Kelso 1995 *Dynamic Patterns* MIT Press

Tap the <u>left</u> index finger
<u>in-phase</u> with the tick
of the metronome.

Try to tap the <u>right</u> index finger
<u>out-of-phase</u> with the tick
of the metronome.

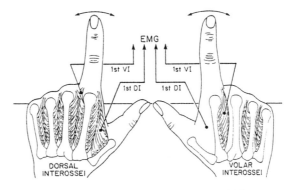

As the frequency of the metronome increases, the right finger shifts
from <u>out-of-phase</u> to <u>in-phase</u> motion.

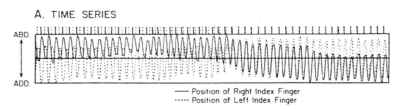

A. TIME SERIES

ABD.

ADD.

— Position of Right Index Finger
----- Position of Left Index Finger

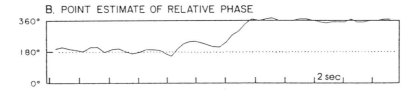

B. POINT ESTIMATE OF RELATIVE PHASE

360°

180°

0°

2 sec

This change can be explained
as a change in a potential
energy function similar to the
change which occurs in a
physical phase transition.

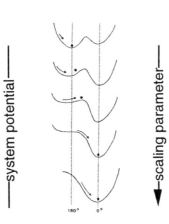

187

Biological Implications of Bifurcations

1. Small Changes in Parameters Can Make Large Changes in Behavior

We are used to thinking that small changes in parameters must produce similarly small changes in the behavior of a system. This intuition is based on our experience with *linear* systems. This behavior is not necessarily true for *nonlinear* systems.

The behavior of a nonlinear system can change dramatically when there is only a small change in the value of a parameter. Such an abrupt change is called a **bifurcation**.

Biological experiments with similar experimental parameters can sometimes produce markedly different results. Biological effects do not always depend smoothly on the values of the experimental parameters. For example, the biological effects of electromagnetic radiation occur within a set of distinct "windows" in the amplitude and frequency parameters of the radiation applied.

Perhaps some of these results arise from the sudden change of behavior when there are small changes in the values of the parameters that characterize bifurcations in nonlinear systems.

2. Bifurcation Sequence to Evaluate Determinism

A biological system can be confirmed to be deterministic if the bifurcation sequence predicted by a deterministic mathematical model of the system matches the bifurcation sequence that is experimentally observed.

3. Bifurcation Sequence versus Phase Space Set

The bifurcation sequence is a qualitative set of behavior changes that may be easier to identify in an experimental system than determining the fractal dimension of the phase space set. Moreover, varying one parameter in an experiment means that the system can serve as its own control.

However, determining the bifurcation sequence requires a good mathematical model of a system that is not needed to use the phase space set analysis.

188

Small changes in parameters can produce large changes in behavior.

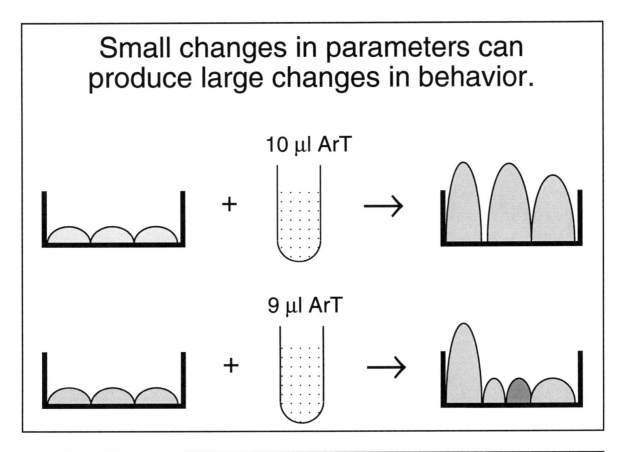

Bifurcations can be used to test if a system is deterministic.

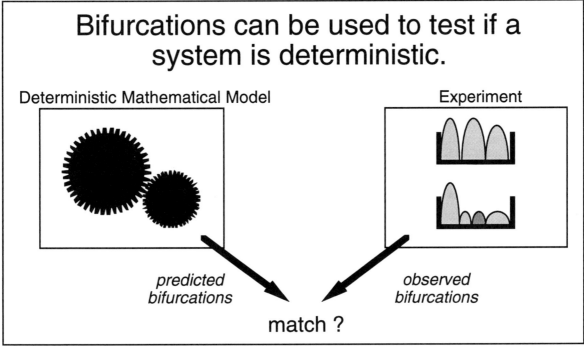

CHAOS
Analyzing Data

Analyzing Data: The Good News

The data from many biological systems have values that seem to change randomly from one point in time to the next point in time, or among repetitions of the same experiment.

It had always been assumed that this variability was generated by random mechanisms based on chance events. We now know that deterministic mechanisms can also generate this variability.

Evaluating the **fractal dimension of the phase space set** or the **bifurcation sequence** now makes it possible to analyze data from an experiment to find out if the variability in the data was generated by a random mechanism or by a deterministic mechanism. If the data were generated by a deterministic mechanism, then we can understand and perhaps even control it.

There has been a lot of excitement in applying these methods to analyze data from many different biological systems to determine if the observed variations were generated by random or deterministic mechanisms.

We have already seen that the low fractal dimension of the phase space sets of the motion of the surface of hair cells in the ear and the timing of the beats of heart cells in culture has demonstrated that these systems are deterministic. We have also seen that the bifurcation sequence has demonstrated that the variation of the concentration of ATP in time produced by glycolysis is deterministic.

Evaluating the dimension of the phase space set has been the method that has been used most to determine if data were generated by a random or a deterministic mechanism. If the **fractal dimension** of the phase space set is very *large*, then the data were generated by a *random* mechanism. If the fractal dimension of the phase space set is *small*, then the data were generated by a *deterministic* mechanism.

The fractal dimension of the phase space set tells us if the data was generated by a random or a deterministic mechanism.

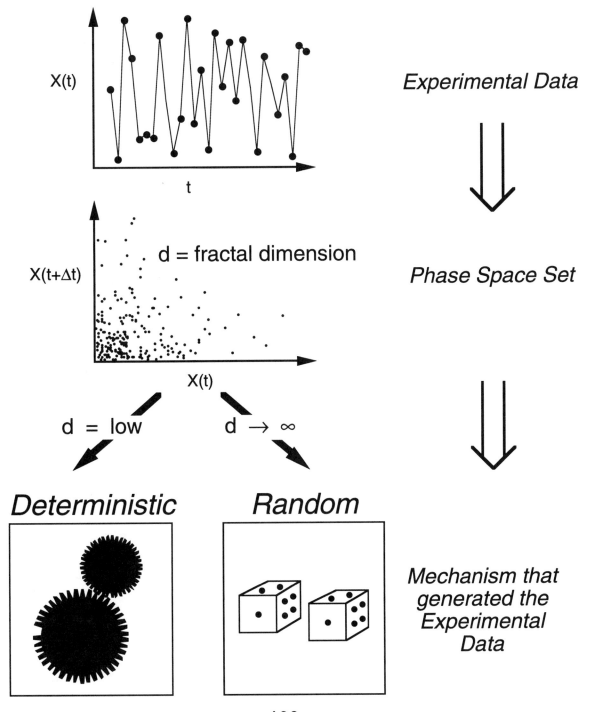

Example of the Fractal Dimension of the
Phase Space Set: Epidemics

There is considerable variation each year in the number of people who get sick with communicable diseases such as **measles** and **chickenpox**. Are these variations due to chance, or are they due to the deterministic characteristics of how these diseases are transmitted from person to person?

Olsen and Schaffer constructed the phase space set from the number of cases reported in time of different diseases in different cities. They found that the fractal dimension of the phase space set depended on the type of the disease and on the size of the city.

The fractal dimension of the phase space set for measles was low. Thus the variations in the number of people with measles was mostly due to a deterministic mechanism. That is, the variations arise from the values associated with the rate of the transmission of the disease from one person to another, the duration of the incubation period before an infected person can transmit the disease, and the duration of the infectious period during which an infected person can transmit the disease.

The fractal dimension of the phase space set for chickenpox was higher than that for measles. The highest fractal dimension that can be found from the phase space set is limited by the amount of data. Given the amount of data that they analyzed, they concluded that the fractal dimension of the phase space set for chickenpox was probably very large. This would mean that the variations in the number of people with chickenpox was due mostly to a random mechanism.

Epidemics

Schaffer and Kot 1986 *Chaos* ed. Holden, Princeton Univ. Press

New York:

	measles	chickenpox
time series:	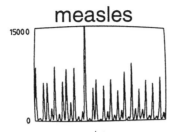	
phase space:		

Olsen and Schaffer 1990 Science 249:499-504

dimension of attractor in phase space

	measles	chickenpox
København	3.1	3.4
Milwaukee	2.6	3.2
St. Louis	2.2	2.7
New York	2.7	3.3

SEIR models – 4 independent variables

S	susceptible
E	exposed, but not yet infectious
I	infectious
R	recovered

Conclusion:

measles: chaotic

chickenpox: noisy yearly cycle

Example of the Fractal Dimension of the Phase Space Set: The Heart

The electrical activity of the heart can be recorded from electrodes placed on the chest. These recordings are called the **electrocardiogram**, which is abbreviated as **ECG** or **EKG**. Each heartbeat generates an electrical pulse. Thus the ECG can be used to measure the time between heartbeats.

Many people are now using "chaos" methods to characterize the normal working of the heart, to differentiate normal and life threatening ECG signals that could be used in automatic clinical monitors, and to find new parameters to predict the future course of heart disease in each person.

A small sample of this work includes constructing phase space sets from either: (1) the ECG voltage as a function of time recorded from the heart, or (2) the intervals of time between consecutive heartbeats.

All the studies have found that there is a *considerable variability in the beating of the normal human heart.*

However, *other results from these studies are less clear.* Different groups have found different fractal dimensions from ECG recordings of the same type of heart condition.

For example, if it is electrically excited, each small piece of the heart will beat on its own. Normally, a wave of electrical activity sweeps across the heart, organizing these separate pieces into a coherent contraction that pumps the blood out of the heart. This organization fails in a condition called *fibrillation.* In fibrillation each piece of the heart beats separately, blood is not pumped, and death follows shortly. Some groups have found that fibrillation is high dimensional, which means that each piece of the heart is beating independently of the other pieces. Other groups have found that fibrillation is low dimensional, which means that the beating of each piece of the heart is actually linked together.

Electrocardiogram

ECG: Electrical recording of the muscle activity of the heart.

time series: voltage

Kaplan and Cohen 1990 Circ. Res. 67:886-892

normal   *fibrillation* → *death*

time series: V(t)

phase space

V(t), V(t+Δt)

D = 1
chaos

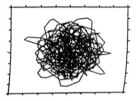

D = ∞

random

Babloyantz and Destexhe 1988
Biol. Cybern. 58:203-211

normal
D = 4
chaos

time series: time between heartbeats

Babloyantz and Destexhe 1988
Biol. Cybern. 58:203-211

normal
D = 6
chaos

Evans, Khan, Garfinkel, Kass,
Albano, and Diamond 1989
Circ. Suppl. 80:II-134

fibrillation → *death*

D = 4
chaos

Zbilut, Mayer-Kress, Sobotka,
O'Toole and Thomas 1989
Biol. Cybern. 61:371-381

induced arrhythmias
D = 3
chaos

Example of the Fractal Dimension of the Phase Space Set: The Brain

The electrical activity of the brain can be recorded from electrodes placed on the head. These recordings are called the **electroencephalogram**, which is abbreviated as **EEG**

Many people are now using chaos methods to characterize the electrical patterns found from the EEG of a person at rest and how those patterns change when the person is learning, remembering, or thinking.

A small sampling of this work includes constructing the phase space set from the EEG voltage recorded as a function of time.

The results of these studies of the EEG of the brain are *not clear.* Different groups have found different fractal dimensions for the phase space set of the EEG under the same conditions.

If there is a consensus, it is that the fractal dimension of the phase space set of the EEG increases with the complexity of an externally imposed task. The lowest fractal dimensions were found from people with severe brain disease, those in the rhythmic convulsions of epilepsy, and those in deep meditation. The fractal dimension increased as the activity level of a person increased from quiet sleep, to a quiet awake state, to performing a mental task (such as counting backwards by sevens).

Electroencephalogram
EEG: Electrical recording of the nerve activity of the brain.

Mayer-Kress and Layne 1987 Ann. N. Y. Acad. Sci.504:62-78

time series: V(t)   **phase space: V(t), V(t+△t)**

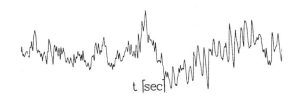

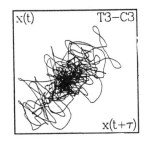

**D = 8
chaos**

Rapp, Bashore, Martinerie, Albano, Zimmerman, and Mees 1989
Brain Topography 2:99-118

Babloyantz and Destexhe 1988 In: From Chemical to
Biological Organization ed. Markus, Müller, and Nicolis, Springer-Verlag

Skarda and Freeman 1987 Behav. Brain Sci. 10:161-195

Xu and Xu 1988 Bull. Math. Biol. 50:559-565

Different groups find <u>different</u> dimensions

under the <u>same</u> experimental conditions.

---

perhaps:

High Dimension ↑ mental task

quiet awake, eyes closed

quiet sleep

brain virus: Creutzfeld-Jakob
epilepsy: petit mal
Low Dimension | meditation: Qi-kong

Ion Channel Kinetics: Random
or Deterministic?

An ion channel protein in the cell membrane switches between states that are open or closed to the flow of ions. We have already seen that the open and closed times have fractal properties. The fractal analysis describes the statistical properties of the open and closed times. It does not tell us whether the switches between the open and closed states are generated by a random or by a deterministic process.

It has been known for 40 years that a random process can generate the statistical properties of the open and closed times found in the experimental data. *We showed that a deterministic process could also generate the same statistical properties.* For example, the following random and deterministic processes each generate the number of times that a channel is open or closed for a duration of time t that is proportional to a single exponential of the form exp(-kt). More complex models can be constructed from pieces consisting of these models that will have numbers of open and closed times proportional to the sum of single exponential terms.

1. Random: Markov Model

There is a certain probability at each time interval that the channel will continue in its present state and a certain probability that it will switch states. Although the probabilities are known, the time at which the switch occurs is set by chance.

2. Deterministic: Iterated Map Model

Let x(n) be the value of the current through the channel at a time n. The value of the current x(n+1) at the next point in time n+1 is a function of the current x(n) at time n. This function can be shown on a plot of x(n+1) versus x(n). In this plot, the y-axis is the value of the current at the next point in time and the x-axis is the value of the current at the present time. One example of such a function is shown by the heavy lines.

To compute the values of the current with time, start with the first value x(1) of the current on the x-axis, move up vertically to the heavy line, and then move left to read off the next value x(2) on the y-axis. Then start with this new value x(2) on the x-axis and repeat the procedure. This type of model is called an **iterative map**.

Random Markov

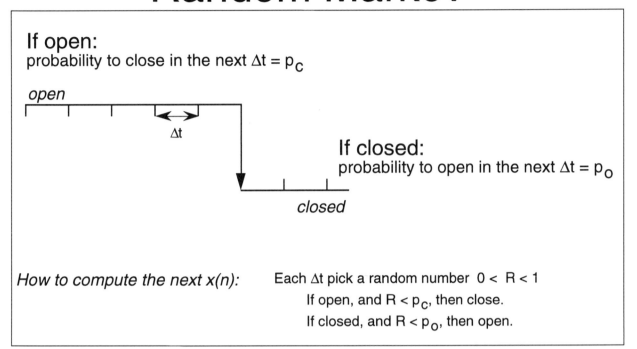

If open:
probability to close in the next $\Delta t = p_C$

open

Δt

If closed:
probability to open in the next $\Delta t = p_O$

closed

How to compute the next x(n):　　Each Δt pick a random number $0 < R < 1$

If open, and $R < p_C$, then close.

If closed, and $R < p_O$, then open.

Deterministic Iterated Map

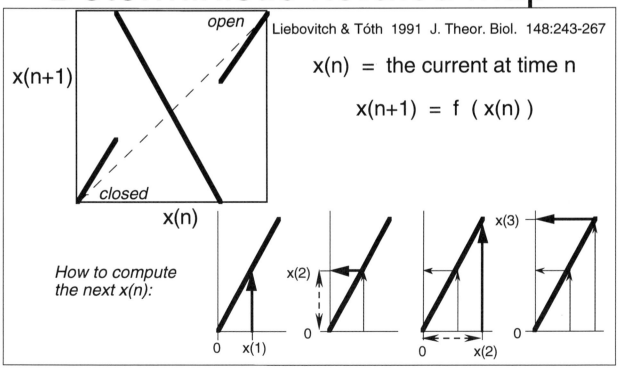

open　　Liebovitch & Tóth　1991　J. Theor. Biol.　148:243-267

$x(n+1)$

$x(n) = $ the current at time n

$x(n+1) = f(x(n))$

closed

$x(n)$

*How to compute
the next x(n):*

$x(2)$　　$x(3)$

0　　$x(1)$　　0　　0　　$x(2)$　　0

Ion Channel Kinetics: An Interpretation
of the Deterministic Model

To breathe some physical life into the abstract mathematics of the deterministic iterative map model, note that the slopes of the lines of the function x(n+1) versus x(n) are greater than 1. This means that increases in the current cause it to increase further and decreases in the current cause it to decrease further. Thus, when the channel is open, a small decrease in the current is soon amplified into further decreases, thus rapidly closing the channel. Similarly, when the channel is closed, a small increase in the current is soon amplified into further increases, thus rapidly opening the channel.

A system that behaved this way, at a larger scale than a channel protein, occurred on Thursday November 7, 1940. It was the failure of the Tacoma Narrows bridge. This failure is often given in college physics courses as an example of simple, forced resonance. This failure was *not* due to simple, forced resonance. In simple, forced resonance you hit on a structure with a rhythm that excites a natural frequency of the structure into motion. The wind that day did not blow rhythmically. Rather, the bridge acting as a nonlinear, mechanical oscillator organized the nonperiodic motion of the wind into periodic motion of the bridge. The motion of the wind caused the bridge to twist in a way that changed the motion of the wind that caused the bridge to twist even more. The wind supplied the energy, but the bridge used that energy to excite itself.

In the same way, an ion channel protein is also a nonlinear, mechanical oscillator made of sticks and springs, which are sometimes called atoms and atomic forces. Perhaps the ion channel can organize the nonperiodic thermal fluctuations in its structure into coherent motions that change its shape from one that is open to one that is closed to the flow of ions.

Tacoma Narrows Bridge

Thursday November 7, 1940

Good modern review (explaining why the explanation given in physics textbooks is wrong):
Billah and Scanlan 1991 Am. J. Phys. 59:118-124

Equation of simple, forced resonance:

$$\ddot{a} + A\dot{a} + Ba = f(t)$$

Equation of <u>flutter</u> that destroyed the Tacoma Narrows Bridge:

$$\ddot{a} + A\dot{a} + Ba = f(a, \dot{a})$$

Wind Tunnel Tests

Scanlan and Vellozzi 1980
in Long Span Bridges
ed. Cohen and Birdsall pp.
247-263 NYAS

The drag on an airplane wing (A) increases with wind speed.

The drag on the OTN (Original Tacoma Narrows) bridge changes sign as the wind speed increases, it enters into positive feedback.

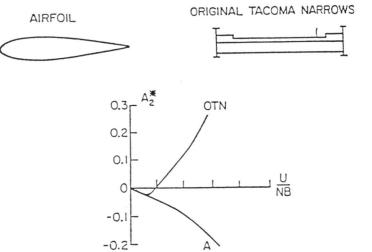

Ion Channel Kinetics: Physical Meaning of the Random and Deterministic Models

1. Random: Driven By Heat

The physical interpretation of the Markov model is that the ion channel protein is like a small molecule that is relentless kicked by thermal fluctuations from one state to another.

It conserves energy and it is therefore in local thermodynamic equilibrium with its environment.

It sits in a state, and sits in a state, until one time, by chance, not based on any physical antecedent, a bit of heat is added from the environment that suddenly provides enough energy for the channel to change its shape from one state to another state.

2. Deterministic: Little Machines with Sticks and Springs

The physical interpretation of the iterated map model is that the ion channel protein is large enough to behave partially as a true mechanical system.

It is a nonlinear, mechanical oscillator composed of sticks and springs. The sticks are called atoms. The springs are called atomic, electrostatic, and hydrophobic forces. This mechanical device organizes nonperiodic thermal fluctuations in its structure into coherent motions. Small motions within the channel protein are amplified by its structure and forces into large, coherent motions that drive the shape of the protein from one state to another.

It dissipates energy and it is therefore not in local thermodynamic equilibrium with its environment.

It is like a miniature, mechanical machine. Its sticks and springs perform reproducible, functionally important motions when set into action by the binding of a ligand or a change in the voltage across the cell membrane.

Random

Like a small molecule, relentlessly kicked by the surrounding heat from one state to another.

The change of states is driven by chance kT thermal fluctuations.

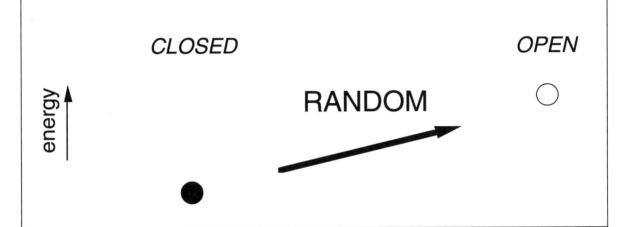

Deterministic

 Like a little mechanical machine with sticks and springs.

The change of states is driven by coherent motions that result from the structure and the atomic, electrostatic, and hydrophobic forces in the channel protein.

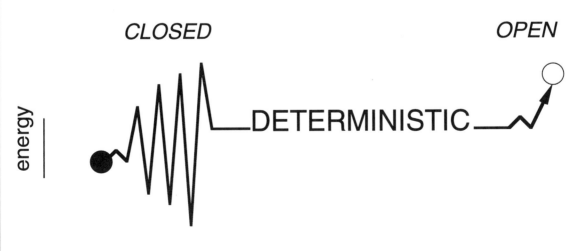

Analyzing Data: The Bad News

We used the patch clamp technique to record the currents through individual ion channels. We then constructed the phase space set from our data to determine if *random,* thermal fluctuations, or *deterministic,* atomic forces cause the channel to switch between its open and closed states. Our conclusion: It was very hard to tell which mechanism caused the switching. Thus we move from the Good News to the Bad News.

1. The Good News

The good news is that these new methods now make it possible to determine if data were generated by a **random** or a **deterministic** mechanism.

2. The Bad News

The bad news is that these methods can be **difficult** and frustrating to use.

The mathematics of chaos is still being developed. New methods may overcome these difficulties.

The present state of affairs should not surprise you or lead you to a low opinion of this field.   A similar situation prevailed four centuries ago when Newton and Leibniz formulated the calculus. The success of the calculus was haunted by definitions that were not clear and results that could not be proved rigorously. The differentials, Δx and Δy used to evaluate the derivative $dy/dx = \Delta y/\Delta x$ were ridiculed as "the ghosts of departed quantities." Afraid of such criticisms, Newton chose to use less controversial geometric arguments, rather than the calculus, in his great book, the *Principia.* It took another 150 years until Cauchy's formulation of the concept of limit put the derivative on firmer ground.

In retrospect, mathematics is clear, rigorous, and obvious. In progress, the crude figures drawn in the sand sometimes look imperfect and unpleasing.

Analyzing Experimental Data

The Good News:

> In principle, you can tell if the data was generated by a <u>random</u> or a <u>deterministic</u> mechanism.

The Bad News:

> In practice, it isn't easy.

Some of the Problems of the Phase Space Analysis

1. How Much Data?

It is not clear how many measurements are needed to construct the phase space set and determine its fractal dimension.

It looks as if a lot of data are needed. It may be difficult to measure this much data from an experiment or to maintain a biological system under constant conditions long enough to record such a large amount of data.

2. The Lag Δt

The phase space set can be constructed from the measurements of one variable $X(t)$. Each point in the N-dimensional phase space then has the coordinates $X(t)$, $X(t+\Delta t)$, $X(t+2\Delta t)$, . . . , $X(t+(N-1)\Delta t)$. It is often not clear what value of the lag Δt should be used to construct the phase space set accurately.

3. The Fractal Dimension of the Phase Space Set

There are different mathematical definitions of the fractal dimension. Any one dimension can be evaluated by different computational methods. The value found for the fractal dimension depends both on the definition and on the method used to evaluate it.

4. Mathematical Limitations of the Embedding Theorems

The mathematics of chaos is still being developed. Some of the existing theorems have assumptions that are not satisfied by the experimental data being analyzed. For example, the embedding theorems require that the sequence of values are "smooth," which is not true if the data are fractal.

Why it's Hard to Tell Random from Deterministic Mechanisms

Need Lots of Data

Very large data sets: 10^d?
Sampling rate must cover the attractor evenly.
   Sample too often: only see 1-d trajectories.
   Sample too rarely: don't see the attractor at all.

Analyzing the Data is Tricky

Choice of lag time Δt for the embedding.
   lag too small: the variable doesn't change enough, derivatives not accurate.
   lag too long: the variable changes too much, derivatives not accurate.
Method of evaluating the dimension.

Mathematics is Not Known

Embedding theorems are only proved for smooth time series.

Problems: How Much Data?

How many values in time are needed in order to accurately construct the phase space set and to evaluate its fractal dimension?

Different groups proposed different methods to estimate the amount of data needed.

Each of these methods is based on a reasonable mathematical argument. The validity of each of these methods was confirmed by computer tests on sample problems done by the group that proposed the method.

To compare the results of the different methods, we used each method to compute the number of values in the time series needed to confirm the existence of a 6-dimensional attractor. The estimates of the number of values needed range from 10 to 5,000,000,000.

This range of uncertainty makes it very difficult to design experiments.

It is not clear how much data is needed. The highest estimates may be too high. However, our experience is that about 10^D measurements are needed for us to identify an attractor of dimension D. This is a lot of data. It raises the question of how a biological system can be maintained under constant experimental conditions long enough to record such a large amount of data.

How Many Time Series Values?

$N = $ Number of values in the time series needed to correctly evaluate the dimension of an attractor of dimension D	N when D=6
Smith 1988 *Phys. Lett.* A133:283 $\quad 42^D$	5,000,000,000
Wolff *et al.* 1985 *Physica* D16:285 $\quad 30^D$	700,000,000
Wolff *et al.* 1985 *Physica* D16:285 $\quad 10^D$	1,000,000
Nerenberg & Essex 1990 *Phys. Rev.* A42:7065 $\quad \frac{1}{K_d^{1/2}[A\ln(k)]^{(D+2)/2}} \frac{D+2}{2} \times [\frac{2(k-1)\Gamma((D+4)/2)}{\Gamma(1/2)\Gamma((D+3)/2)}]^{D/2}$	200,000
Ding *et al.* 1993 *Phys. Rev. Lett.* 70:3872 $\quad 10^{D/2}$	1,000
Gershenfeld 1990 preprint $\quad \dfrac{2^D (D/2)!}{\pi^{D/2}}$	10

Problems: The Lag Δt

When the phase space set is constructed from the time series of one measured variable $X(t)$, then the coordinates of each point in the N-dimensional phase space are $X(t)$, $X(t+\Delta t)$, $X(t+2\Delta t)$, . . . , $X(t+(N-1)\Delta t)$. The properties of the phase space set depend on the value used for the lag Δt. This can be seen by constructing the 2-dimensional phase space from the variable $X(t)$ of the Lorenz system.

When the lag **Δt is too small**, then the value of $X(t+\Delta t)$ is almost the same as the value of $X(t)$. Thus the phase space set lies along the diagonal $X(t+\Delta t) = X(t)$. The phase space set is 1-dimensional. Since the **fractal dimension** of the phase space set is **low**, you would conclude that the time series $X(t)$ was generated by a **deterministic** mechanism.

When the lag **Δt is too large**, because of the sensitivity to initial conditions, the value of $X(t+\Delta t)$ is uncorrelated with the value of $X(t)$. Thus the phase space set fills the phase space and is therefore high dimensional. Since, the fractal dimension of the phase space set is **high**, you would conclude that the time series $X(t)$ was generated by a **random** mechanism.

It is only when the lag **Δt is just right**, that the **phase space set** has the **real** form of the attractor of the Lorenz system and the real fractal dimension of approximately 2.03.

There are a number of excellent methods to determine Δt in principle. These methods are based on finding the time scale Δt of the correlations between $X(t)$ and $X(t+\Delta t)$ using the autocorrelation or mutual information functions. These methods were developed and tested on data generated by differential equations where there is a unique correlation time scale. However, in practice, experimental data often have more than one correlation time scale. The fractal dimension of the phase space set will depend on which time scale is used for the lag Δt. Different values of the lag will produce phase space sets with different fractal dimensions. There is no way to choose which is the "real" time scale and its associated lag Δt, that gives the "real" value of the fractal dimension of the phase space set.

Lorenz

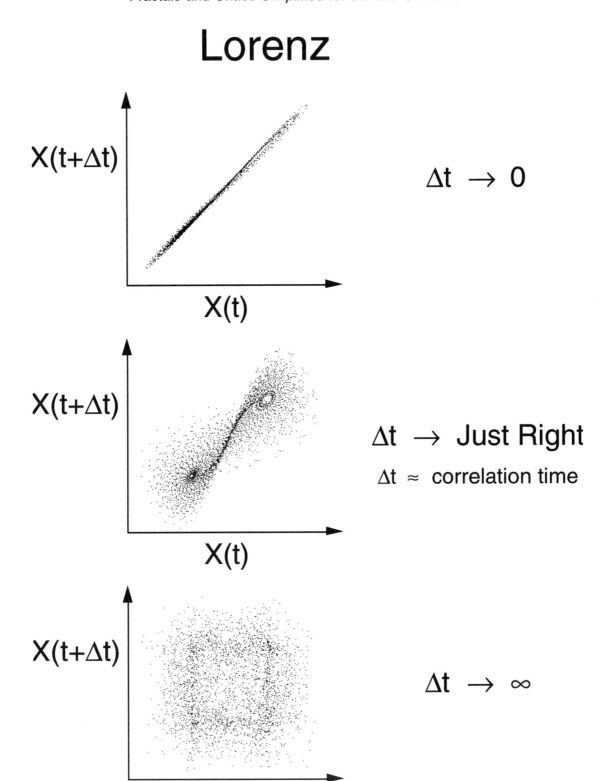

Problems: The Embedding Theorems

We can always construct a plot from the points with coordinates X(t), X(t+Δt), X(t+2Δt), . . . , X(t+(N-1)Δt), where X(t) is the time series of one variable. Takens' theorem tells us that if certain conditions are met, then this plot is related to the phase space set. For example, the theorem assumes that the time series X(t) is "smooth," that is, that the derivatives dX(t)/dx and $d^2X(t)/dt^2$ exist.

However, in practice, experimental data may not have the properties assumed by Takens' theorem. For example, when the time series X(t) is fractal, then the finer we look in time, the more wiggles we find. Thus the derivatives dX(t)/dx and $d^2X(t)/dt^2$ do *not* exist. We can still plot the points with coordinates X(t), X(t+Δt), X(t+2Δt), . . . , X(t+(N-1)Δt), but because the data do not satisfy the assumptions of the theorem, we do not know how, if at all, this plot is related to the "real" phase space set.

Osborne and Provenzale showed that this embedding procedure does not produce the real phase space set from a time series of fractal data. They constructed the lag plot of points with coordinates X(t), X(t+Δt), X(t+2Δt), . . . , X(t+(N-1)Δt). Since their time series was generated by a random mechanism, the fractal dimension of the real phase space was infinite. However, they found that the fractal dimension of the lag plot was sometimes as low as 1, which is a lot less than infinity. Thus, for this type of data, the lag plot is not the "real" phase space set.

Experimental data do not always satisfy the assumptions of the embedding theorems. **It would be very worthwhile for mathematicians to formulate and prove theorems that would tell us how to construct the phase space set from data that are not continuous or differentiable.**

Takens' Theorem

If

The time series X(t) is "smooth":

$$\frac{dX(t)}{dt} \quad \frac{d^2X(t)}{dt^2} \quad \text{exist}$$

+ other stuff too

Then,
*the **lag plot** constructed from the data*

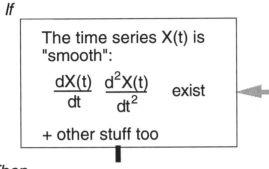

is a linear transformation of
*the **real phase space***

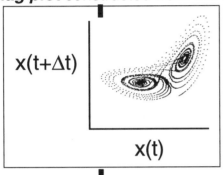

because $\quad \dfrac{dX(t)}{dt} \approx \dfrac{X(t+\Delta t) - X(t)}{\Delta t}$

Since the fractal dimension is
invariant under a linear transformation,
*the fractal dimension of the **lag plot***
is equal to the fractal dimension
*of the **real phase space set**.*

If the data does **not** satisfy these assumptions then we are **not** guaranteed that the fractal dimension of the **lag plot** is equal to the fractal dimension of the **real phase space set**.

For example:

The ion channel current is **not** smooth, it is fractal (bursts within bursts) and therefore <u>NOT</u> differentiable.

Thus, the assumptions of the theorem are <u>not</u> met and we are <u>not guaranteed</u> that the fractal dimension of the **lag plot** is equal to the fractal dimension of the **real phase space set**.

For example:

Osborne & Provenzale. 1989.
Physica D35:357-381.

They used a Fourier series to generate a fractal time series whose power spectra was $1/f^{\alpha}$. They randomized the phases of the terms in the Fourier series so that the fractal dimension of the real phase space set was <u>infinite</u>. But, they found that the fractal dimension of the lag plots was as low as <u>1</u>.

Problems: The Meaning of a
Low Dimensional Attractor

It is always hard in science to prove that *an* explanation is *the* explanation. The existence of a low dimensional attractor means that a deterministic process could have generated all the values of the measurements found in the experiment. It does not mean that the deterministic process did generate these values.

For example, let's say we have a set of a million numbers, each of which happens to be 6. We pick numbers from this set at **random**. The time series will be: 6, 6, 6, 6, 6, ... The phase space set will be a zero-dimensional point. The low fractal dimension of the phase space set accurately indicates that a **deterministic** mechanism can generate these data. For example, the deterministic mechanism could be: (1) let the value of the first number be equal to 6, and (2) let the value of each subsequent number be equal to the value of the previous one. The problem is that even though a deterministic mechanism *could* have generated the data, in this case it didn't. These data were generated by a random mechanism.

We are more likely to believe in the uniqueness of an explanation in science when that explanation predicts new values, explains previously unrelated facts, or allows us to see things in a new light that helps us make further discoveries.

Pathological example where an
infinite dimensional random process
has a **LOW** dimensional attractor

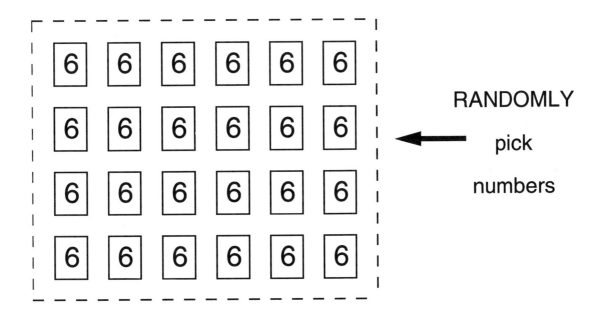

RANDOMLY

pick

numbers

Time Series: 6, 6, 6, 6, 6, 6, 6, 6 . . .

Phase Space:

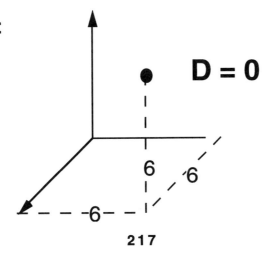

D = 0

217

New Methods: Average Direction
of the Trajectories

There is an ongoing effort to overcome the problems in the phase space analysis and to develop new methods.

As a system evolves in time, the point that represents the state of the system moves through the phase space. The trail of this point through the phase space is called the **trajectory**. The trajectory forms the phase space set. A large amount of data is needed to produce a long enough trajectory to determine the fractal dimension of the phase space set. Instead of evaluating the fractal dimension of the phase space set, Kaplan and Glass suggested evaluating the motions through the phase space set. This can be done by evaluating the **average directions** of the trajectories that pass through each small region in the phase space.

1. Random: Average Directional Vector Is Small

When the trajectory is constructed from data that were generated by a random mechanism, then the motion of the point in phase space is an aimless random walk. Each time the point passes through a small region in phase space it will pass through going in a different direction. Thus the average of the directions will be **small**, because the motion in different directions will cancel out.

2. Deterministic: Average Directional Vector Is Large

When the trajectory is constructed from data that were generated by a deterministic mechanism, then the motion of the point in phase space is highly organized, forming an attractor. Each time the point passes through a small region in phase space it will pass through going in approximately the same direction. Thus the average of the directions will be **large**, because the motions are in the same direction and so add together.

Organization of the Vectors in the Phase Space Set

Kaplan and Glass 1992 Phys. Rev. Lett. 68:427-430

Random

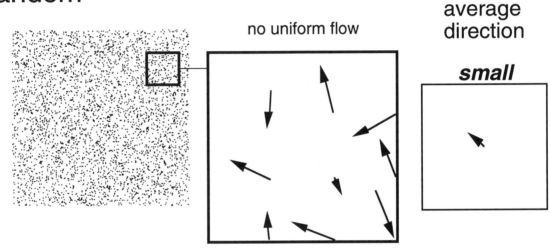

no uniform flow

average direction

small

Deterministic

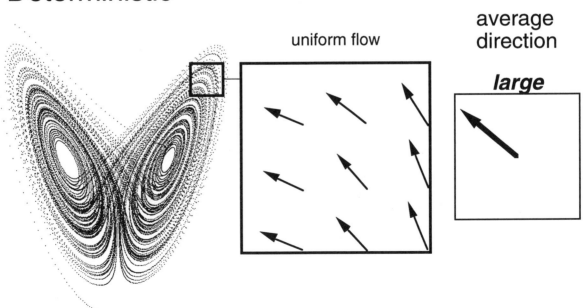

uniform flow

average direction

large

New Methods: Surrogate Data Set

The errors in evaluating the fractal dimension of the phase space set should be similar when similar data sets are analyzed. Instead of trying to determine the absolute value of the fractal dimension, we can determine how it changes when we change the data. For example, we can scramble the original data to remove any deterministic relationship that might be present. The scrambled version of the original time series is called the **surrogate data set**. If we find that the dimension of the phase space set of the surrogate data set is different from that of the original data, then that difference must be due to the fact that we removed a deterministic relationship that was present in the original data.

The original data can be scrambled in different ways to generate the surrogate data set. For example, the surrogate data set can consist of the same values of the original time series in a random order. This scrambling may be too severe because consecutive values of the surrogate data set are uncorrelated. Theiler suggested that a better surrogate data set might be one that has the same first order correlations as those in the original time series but where all the higher order correlations have been removed. This can be generated by randomizing the phases of the Fourier components of the original time series.

1. Random: Original and Its Surrogate Are the Same

A surrogate time series with the same first order correlations but no higher order correlations is generated from the original time series. If the fractal dimension of the phase space set of the original time series is the same as that of its surrogate, then no deterministic relationship was present in the original time series.

2. Deterministic: Original and Its Surrogate Are Different

If the fractal dimension of the phase space set of the original time series is less than that of its surrogate, then the scrambling used to generate the surrogate removed a deterministic relationship that was present in the original time series.

Surrogate Data Set
Theiler et al. 1992 Physica D58:77-94

Random

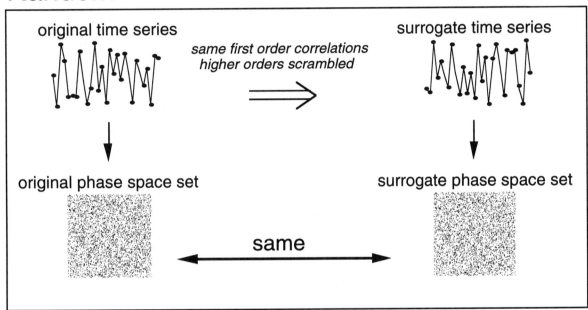

Deterministic

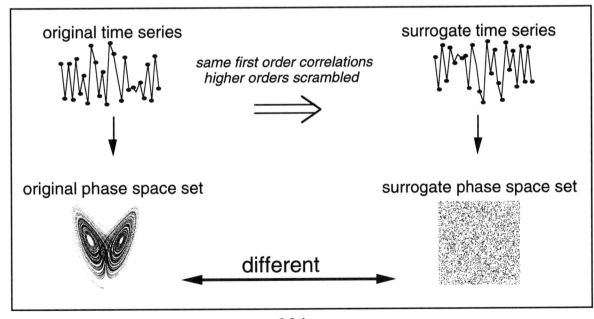

Biological Implications of the Strengths and Weakness in Analyzing Data

1. Methods of Analysis

Some biological data look as if they were generated by random mechanisms. We now know that deterministic, chaotic mechanisms can also generate data that look as if they were generated by random mechanisms. **There are now methods to analyze biological data to determine if they were generated by a random or a deterministic mechanism.** If the data were generated by a deterministic mechanism, then we may be able to understand and even control that biological process.

We have shown that measures that can tell us if the data were generated by a random or a deterministic process are: the *fractal dimension of the phase space set*, the *bifurcation sequence*, the *average trajectories* in small regions of the phase space, and the use of *surrogate data sets*. Additional measures include: the *Liapunov exponents* (how fast trajectories separate from each other in phase space and thus a direct measure of the sensitivity to initial conditions), *nonlinear prediction* (a deterministic relationship can predict subsequent values in the time series), and the *kneading sequence of the symbolic dynamics* (deterministic relationships in the trajectories).

2. Experimental Considerations

Many of these methods of analysis require lots of data. It is not clear how to maintain a living, biological system under constant conditions long enough to record this amount of data.

Because of the problems in using these methods to analyze data, it is best if **the experimental system can serve as its own control**, that is, to design the experiment with a control parameter so that the behavior of the system, such as chaotic or non-chaotic, depends on the value of the control parameter. The results of the analysis of the different behaviors can then be compared to each other. This is a more reliable way to identify a chaotic system than one uncontrolled measurement of the fractal dimension of the phase space set.

Experiments

Weak

Time Series ⟶ phase space

⟶ Dimension

Low = deterministic
High = random

---

examples: **ECG,   EEG**

Strong

vary a parameter ⟶ see behavior

predicted by a
nonlinear model

---

examples: **electrical stimulation of cells,**
biochemical reactions

CHAOS

Control of Chaos

The Advantages of Chaos

We are often impressed by the exquisite detail and sophisticated function of biological systems. Thus we might think that the complex and unpredictable output of a chaotic system would be useless or even dangerous to living things. In fact, chaotic systems can provide important characteristics needed by biological systems.

1. Variability

The variability in the output generated by a chaotic system means that the variables of the system will span a wider range of values. This variability can be useful to biological systems.

For example, in learning, chaotic processes in the brain could explore a wide range of new neural connections to encode the new information. In remembering, chaotic processes could search a wide range of memories to find the goal of the search. Chaotic processes in the heart could increase the variability of the heart rate to help it adjust to changing internal and external demands.

A lack of variability may characterize an unhealthy state. For example, the rhythmic seizures in epilepsy or some of the rhythmic patterns of the beating of a diseased heart represent a lack of variability.

2. Control

Paradoxically, the most surprising advantage of chaos is that it allows both **finer** and **faster control** of systems.

In linear systems, small changes in the input produce small changes in the output. To produce a big change in the output therefore requires a big change in the input.

However, chaotic systems have sensitivity to initial conditions and bifurcations in behavior with small changes in parameters. A delicate adjustment of the input can make a dramatic change in the output.

Thus chaotic systems can be controlled finer and faster, with less energy, than linear systems.

Control

Non-Chaotic System

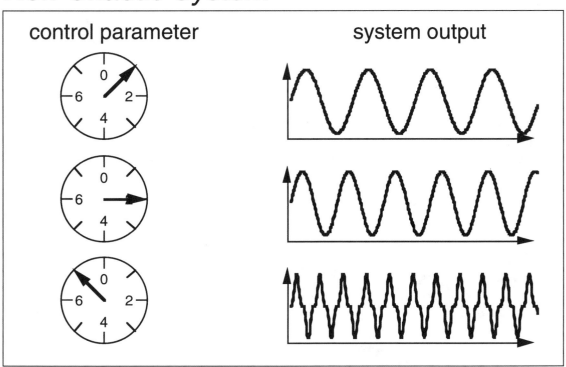

Chaotic System

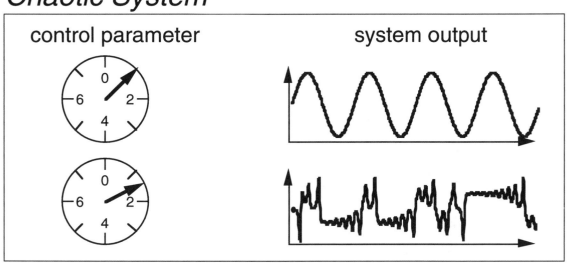

Example of Chaotic Control:
The Light Output from a Laser

Recently, mathematicians have shown how to use small external inputs to control a chaotic system. These new methods have now been successfully used to control electronic and mechanical systems.

For example, the physical process that generates the light in a laser is chaotic. The intensity of the light produced by the laser varies in time. The output of light depends on the values of control parameters. Roy et al. computed how the values of these control parameters should be changed in time in order to control the output of a laser.

They used these mathematical predictions to control the light output of a laser by changing the values of the control parameters with time. Without the control input, the intensity of the light output of the laser varied irregularly in time. With the control input, the intensity of the light changed in a regular, periodic way. By changing the control input, they could vary the duration of this period. Only small changes in the control parameters were needed for this control.

Using only small control inputs they were able to achieve highly accurate control of the light output of the laser.

Control of Chaos

light intensity of a laser

Roy et al. 1992 Phys. Rev. Lett. 68:1259-1262

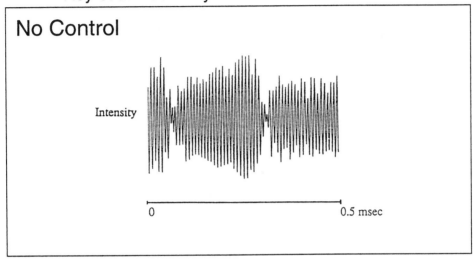

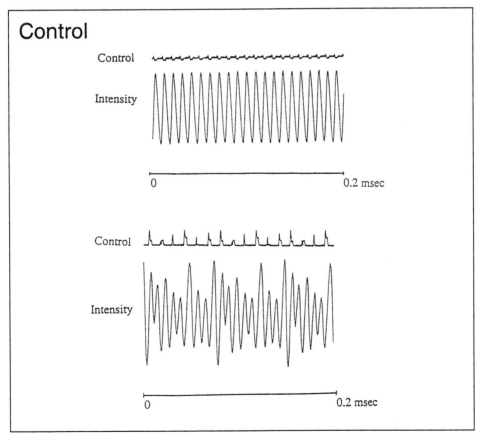

Example of Chaotic Control:
The Motion of a Magnetoelastic Ribbon

These new methods of the control of chaos have also been used to control mechanical systems.

Objects can be fabricated that have a stiffness that depends on the surrounding magnetic field. A magnetoelastic ribbon looks like a piece of audio recording tape. It is quite flexible. If it is supported only on the bottom, then it bends over. However, when placed in a magnetic field, its stiffness increases, and it stands upright.

Ditto et al. put a magnetoelastic ribbon in a magnetic field generated by an electromagnet. They varied the magnetic field in a periodic way, by varying the voltage into the electromagnet. They measured the distance X_n of the ribbon from a sensor at the same time during the n-th cycle of the magnetic field.

They used the control of chaos methods to predict how small additional voltages should be applied to the electromagnet to control the motion of the ribbon. When there was no control, the motion of the ribbon was chaotic. That is, the distance X_n of the ribbon from the sensor changed with subsequent values of n. This appears as a smear on the plot of X_n versus n. When the control was applied, then the motion of the ribbon became periodic. That is, the distance X_n of the ribbon from the sensor had the same value at each cycle of the magnetic field. These repeated values form a horizontal line on the plot of X_n versus n. They could also change the control to double the period of the ribbon. That is, the distance X_n of the ribbon from the sensor had one value on one cycle of the magnetic field and a different value on the next cycle, These two values repeated on alternating cycles. These two values form two horizontal lines on the plot of X_n versus n.

The control took effect very fast when it was turned on, and it was highly accurate.

Control of Chaos

motion of a magnetoelastic ribbon

Ditto, Rauseo, and Spano 1990 Phys. Rev. Lett. 65:3211-3214

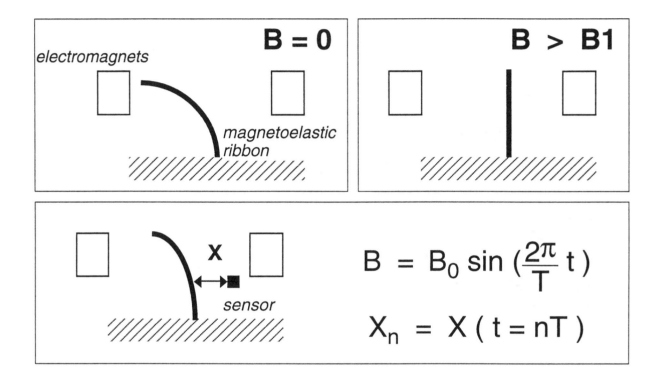

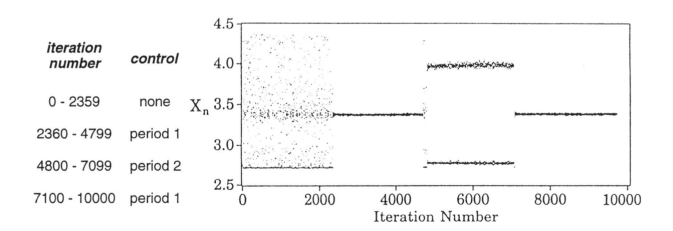

iteration number	control
0 - 2359	none
2360 - 4799	period 1
4800 - 7099	period 2
7100 - 10000	period 1

Biological Implications of the Control of Chaos

1. Clinical Control

We used to think that the variability of biological systems is generated by mechanisms that we will never be able to understand because it involves chance events. However, if the variability of some biological systems arises from deterministic chaos, then we may be able to understand and even **control** these systems.

New mathematical methods have been developed to control chaotic systems. Some of the people who have used these methods to control electronic and mechanical systems have also used them to control the beating of a piece of the heart and the electrical activity in a piece of the brain.

The success of these early experiments suggests the tantalizing promise that these methods may lead to clinical applications where small electrical signals are used to control abnormal rhythms in the heart or seizures in the brain. More speculatively, these methods could be used to control the level of substances in the blood, such as glucose, or the number of cells of different types in the immune system.

Control of Biological Systems

The Old Way

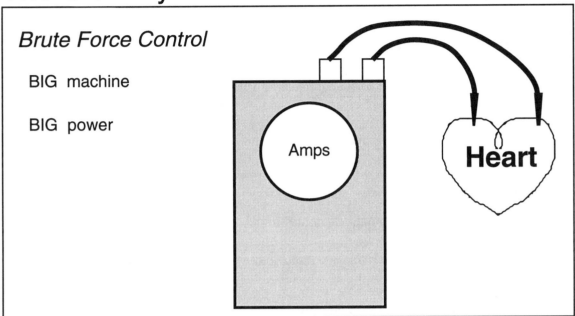

The New Way

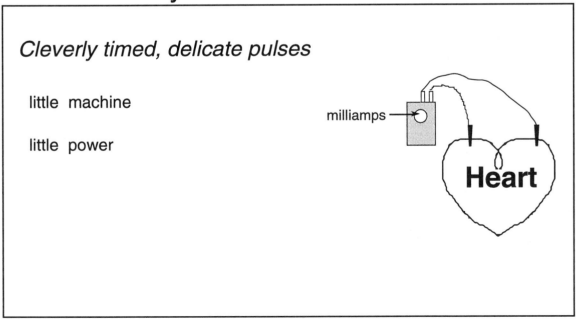

Biological Implications of the Control of Chaos (continued)

2. Are Biological Systems Stable or Unstable?

We do not understand enough about biology to know how to think about living things. Thus we interpret the results of biological experiments using concepts from machines that we build and understand.

Our technology has strived to build things that are stable. Stable systems are tolerant to small errors in control or small changes in the environment. An airplane continues to fly when the hand eases on the control or there is an additional gust of wind. We have thought of biological systems as also having these properties. That is, that biological systems are stable and that they are always trying to keep themselves stable. This concept is called *homeostatis.*

But stability means that a system is hard to control. The stable airplane wants to fly straight, and doesn't want to turn. In order to make an airplane more maneuverable, we must make it *unstable*, and then *control* it. This means that the wires and the computer must work or the airplane will crash.

Now that we have faster ways to process information, we are beginning to build systems that are inherently unstable and then control them. Unstable systems, such as **chaotic systems**, can be **controlled finer** and **faster** than stable systems. Perhaps living things have also chosen to use unstable systems to achieve better control.

In the Lorenz system, cylinders of air rotate clockwise or counterclockwise. The old way of biological thinking would be to interpret each direction of rotation as a *stable state* and that a *force* then switches the rotation of the system. This is not what happens in the Lorenz system. Both directions of rotation are unstable. The act of rotating in one direction causes the switching to the other direction of rotation.

This suggests a new way to think about biological systems. Instead of a biological system switching between stable homeostatic states, perhaps we need to think more of how **the dynamics of being in one condition itself causes the system to switch to another condition**.

How do we think of biological systems?

The Old Way

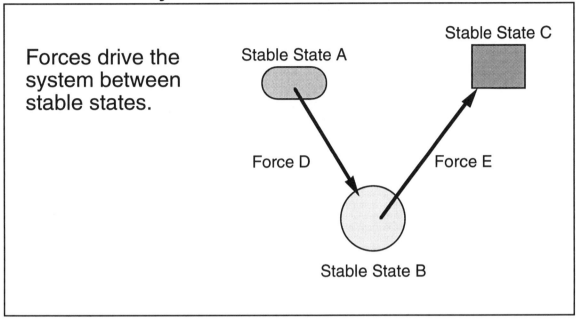

Forces drive the system between stable states.

Stable State A

Stable State C

Force D

Force E

Stable State B

The New Way

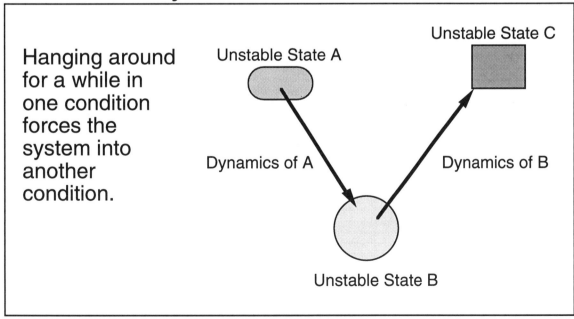

Hanging around for a while in one condition forces the system into another condition.

Unstable State A

Unstable State C

Dynamics of A

Dynamics of B

Unstable State B

CHAOS
Summary

Summary of Chaos

1. Deterministic System with Complex Behavior

A chaotic system is **deterministic**. It can be described by a **small number of independent variables**. However, the output of the system is so **complex** that it mimics the behavior of a random system based on chance. The word chaos was chosen to describe this complex behavior. It does *not* mean that a chaotic system is driven by disorder, randomness, or chance.

2. Dynamical System

Deterministic means that the subsequent values of the variables can be computed from their previous values. This is called a **dynamical system**.

3. Sensitivity to Initial Conditions

The subsequent values of the variables computed depend sensitively on the exact starting values of the variables. If the computation is rerun with slightly different starting values, then the final values of the variables will be significantly different. The values of the variables at each small time step into the future can be computed from their present values. However, the values of the variables far into the distant future cannot be computed from their present values. Thus, the system is **deterministic** over each small time step, but it is **not predictable** in the long run.

4. Phase Space

The values of the variables in time can be transformed into an object in space. The space is called the phase space. The object is called the **phase space set**. *It is easier to determine some properties of the system by analyzing the spatial properties of the phase space set rather than the temporal properties of the values of the variables.* For example, if the **fractal dimension** of the phase set is **high**, then the values of the variables in time were generated by a **random** mechanism. If the fractal dimension of the phase space set is **low**, then the values of the variables in time were generated by a **deterministic** mechanism.

Summary of Chaos

FEW INDEPENDENT VARIABLES

Whose behavior is so complex that it
mimics random behavior.

DYNAMICAL SYSTEM

DETERMINISTIC

The value of the variables at the next instant in time can be
calculated from their values at the previous instant in time.

$$x_i(t+\Delta t) \;=\; f(\,x_i(t)\,)$$

SENSITIVITY TO INITIAL CONDITIONS

NOT PREDICTABLE IN THE LONG RUN

$$x_1(t+\Delta t) - x_2(t+\Delta t) \;=\; A\,e^{\lambda\,\Delta t}$$

STRANGE ATTRACTOR

Phase space is low dimensional (often fractal).

Where to Learn More about Chaos

Chaos is one subject area within the field of **nonlinear dynamics,** which is part of the broader field of **dynamical systems.** There are many good books on chaos and hundreds of research articles published in journals each year. Some references, at different levels, that can lead you further into the mathematical details and the applications of chaos are the following:

1. Introductory Level

The best selling book *Chaos: Making a New Science* by Gleick introduced the concept of chaos to many scientists and nonscientists. It provides a clear and accurate account of the basic ideas without the use of mathematics.

2. Intermediate Mathematical Level

Chaotic and Fractal Dynamics by Moon provides a good introduction into the concepts of dynamic systems and their use in analyzing experimental data. The book shows the relevance of these concepts in many real life engineering applications. The mathematical level is that of ordinary differential equations.

3. Advanced Mathematical Level

Nonlinear Oscillations, Dynamical Systems, and Bifurcations of Vector Fields by Guckenheimer and Holmes provides a mathematically rigorous introduction to dynamical systems. *Chaos* by Ott presents a mathematically rigorous introduction to chaotic systems and current research topics in a contemporary style.

4. Chaos in Biomedical Research

A range of different biomedical applications are described by articles in the review *Chaos,* edited by Holden, and *Complexity, Chaos, and Biological Evolution,* edited by the Moskildes. *Fractal Physiology* by Bassingthwaighte, Liebovitch, and West provides an introduction to chaos at both a qualitative level and at the mathematical level of elementary calculus. It gives detailed descriptions and references of many biomedical applications of chaos.

Books About Chaos

introductory

J. Gleick
Chaos: Making a New Science 1987 Viking

intermediate mathematics

F. C. Moon
Chaotic and Fractal Dynamics 1992 John Wiley & Sons

advanced mathematics

J. Guckenheimer & P. Holmes
Nonlinear Oscillations, Dynamical Systems, and Bifurcations of Vector Fields 1983 Springer-Verlag
E. Ott
Chaos in Dynamical Systems 1993 Cambridge Univ. Press

reviews of chaos in biology

A. V. Holden
Chaos 1986 Princeton Univ. Press
E. & L. Moskilde
Complexity, Chaos, and Biological Evolution 1991 Plenum
J. Bassingthwaighte, L. Liebovitch, & B. West
Fractal Physiology 1994 Oxford Univ. Press

Part III
OTHER
METHODS

Fractals and chaos are only two examples
of a large number of nonlinear methods.

OTHER METHODS

The Big Picture

The Big Picture

Four centuries ago, Newton showed that mathematical descriptions give us insight into the nature of things. However, our mathematics has been mostly limited to simple systems with linear interactions. This corresponds to systems with **few pieces** that **do not interact strongly** with each other.

Since we have not had the tools to do anything else, we analyzed the world as if it consisted of systems with these properties. But the world, especially the world of living things, is not at all like this. The living world is filled with systems that have **many pieces** that **interact strongly** with each other.

Only now are we beginning to develop the appropriate tools to analyze the real world of systems with many strongly interacting pieces. These tools give us a dictionary of examples of what to expect from our experiments and how we should think about our systems. Our vision is only as broad as the different examples that we have assembled in our dictionary.

The World

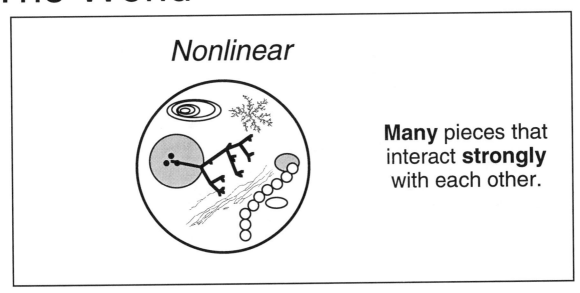

Nonlinear

Many pieces that interact **strongly** with each other.

Most of Our Mathematical Tools

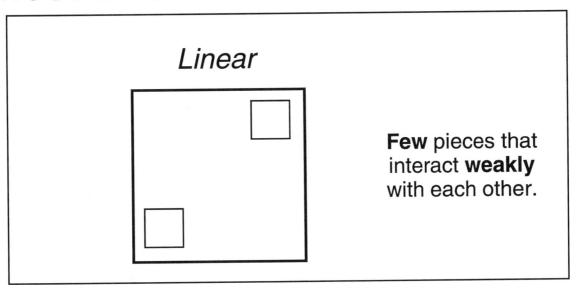

Linear

Few pieces that interact **weakly** with each other.

Some Other Nonlinear Tools

Fractals and chaos are only two tools from a wider class of nonlinear tools now being developed to help us understand systems that have **many pieces** that **interact strongly** with each other. Examples of some additional tools are:

1. Self-Organizing Critical Systems

These systems are deterministic systems with a large number of independent variables, unlike chaos, which has a small number of independent variables. These systems live just at the border of stability. If pushed over the edge, they relax back to just over the stable border. Thus these systems self-organize to live at the stable-unstable transition. A system poised at such a phase transition is called a critical system. These systems can generate fractals in space and time.

2. Neural Networks

These systems consist of nodes that have values and connections between them. At each computational step, the new value of a node depends on the values of the other nodes and on the strengths of the connections between them. Patterns of the values of the nodes are called memories. A set of starting values of the nodes evolve to that of the closest memory. Using rules to change the values of the connection strengths, these systems can learn new memories.

3. Cellular Automata

These systems consist of a set of boxes, each of which is in one of a small number of different possible states. At each computational step, a rule determines the new states of the boxes from the states of the surrounding boxes.

4. Coupled Maps

In these systems each point of an array is a chaotic system. Each chaotic system evolves in time and interacts with the surrounding chaotic systems.

248

Self-Organizing Critical Systems

Bak and Creutz 1994 In Fractals in Science, ed. Bunde and Havlin, pp. 26-47

deterministic, high dimensional

when stress exceeds critical value,
spread it to neighboring points

e.g. sandpile

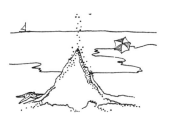

Neural Networks

Amit 1989 Modeling Brain Function Cambridge Univ. Press.

memories = sets of values of the nodes

can find a stored memory closest to an input pattern

can learn new memories

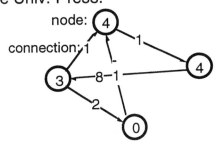

Cellular Automata

Wolfram 1994 Cellular Automata and Complexity Addison-Wesley

can compute

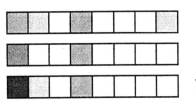

time

Coupled Maps

Kaneko 1993 Physica D68:299

chaotic systems coupled on a grid

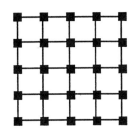

References

D. J. Amit. 1989. *Modeling Brain Function.* Cambridge University Press, New York.

D. Avnir, ed. 1989. *The Fractal Approach to Heterogeneous Chemistry.* John Wiley & Sons, New York.

A. Babloyantz and A. Destexhe. 1988. Is the normal heart a periodic oscillator? *Biol. Cybern.* 58:203-211.

A. Babloyantz and A. Destexhe. 1988. The Creutzfeld-Jakob disease in the hierarchy of chaotic attractors. In *From Chemical to Biological Organization,* M. Markus, S. Muller, and G. Nicolis, eds. Springer-Verlag, New York, pp. 307-316.

P. Bak and M. Creutz. 1994. Fractals and self-organized criticality. In *Fractals in Science,* A. Bunde and S. Havlin, eds. Springer-Verlag, New York, pp. 26-47.

P, Bak, C. Tang, and K. Wiesenfeld. 1988. Scale invariant spatial and temporal fluctuations in complex systems. In *Random Fluctuations and Pattern Growth,* H. E. Stanley and N. Ostrowsky, eds. Kluwer Academic Pub., Boston, pp. 329-335.

M. Barnsley. 1988. *Fractals Everywhere.* Academic Press, New York.

J. B. Bassingthwaighte, L. S. Liebovitch, and B. J. West. 1994. *Fractal Physiology.* Oxford University Press, New York.

J. B. Bassingthwaighte and J. H. G. M. van Beek. 1988. Lightning and the heart: Fractal behavior in cardiac function. *Proc. IEEE* 76:693-699.

K. Y. Billah and R. H. Scanlan. 1991. Resonance, Tacoma Narrows bridge failure, and undergraduate physics textbooks. *Am. J. Phys.* 59:118-124.

A. Block, W. von Bloh, and H. J. Schellnhuber. 1990. Efficient box-counting determination of generalized fractal dimensions. *Phys. Rev. A* 42:1869-1874.

L. M. Boxt, J. Katz, F. C. Czegledy, L. S. Liebovitch, R. Jones, P. D. Esser, and L. M. Reid. 1994. Fractal analysis of pulmonary arteries: The fractal dimension is lower in pulmonary hypertension. *J. Thoracic Imaging* 9:8-13.

S. R. Brown and C. H. Scholz. 1985. Broad bandwidth study of the topography of natural rock surfaces. *J. Geophys. Res.* 90:12575-12582.

J. Cairns, J. Overbaugh, and S. Miller. 1988. The origin of mutants. *Nature* 335:142-145.

E. B. Cargill, H. H. Barrett, R. D. Fiete, M. Ker, D. D. Patton, and G. W.

Seeley. 1988. Fractal physiology and nuclear medicine scans. *SPIE Medical Imaging II* 914:355-361.

F. Caserta, H. E. Stanley, W. D. Eldred, G. Daccord, R. E. Hausman, and J. Nittmann. 1990. Physical mechanisms underlying neurite outgrowth: A quantitative analysis of neuronal shape. *Phys. Rev. Lett.* 64:95-98.

A. M. Churilla, W. A. Gottschalke, L. S. Liebovitch, L. Y. Selector, A. T. Todorov, and S. Yeandle. 1996. Membrane potential fluctuations of human T-lymphocytes have fractal characteristics of fractional Brownian motion. *Ann. Biomed. Engr.* 24:99-108.

M. Ding, C. Grebogi, E. Ott, T. Sauer, and J. A. Yorke. 1993. Plateau onset for correlation dimension: When does it occur? *Phys. Rev. Lett.* 70:3872-3875.

W. L. Ditto, S. N. Rauseo, and M. L. Spano. 1990. Experimental control of chaos. *Phys. Rev. Lett.* 65:3211-3214.

G. A. Edgar. 1990. *Measure, Topology, and Fractal Geometry.* Springer-Verlag, New York.

S. J. Evans, S. S. Khan, A. Garfinkel, R. M. Kass, A. Albano, and G. A. Diamond. 1989. Is ventricular fibrillation random or chaotic? *Circulation Suppl.* 80:II-134.

F. Family, B. R. Masters, and D. E. Platt. 1989. Fractal pattern formation in human retinal vessels. *Physica* D38:98-103.

J. Feder. 1988. *Fractals.* Plenum, New York.

W. Feller. 1968. *An Introduction to Probability Theory and Its Applications,* Vol. 1, 3rd. Ed. John Wiley & Sons, New York.

N. A. Gershenfeld. 1990. On measuring large dynamical dimensions. Preprint.

I. Giaever and C. R. Keese. 1989. Fractal motion of mammalian cells. *Physica* D38:128-133.

L. Glass, M. R. Guevara, J. Bélair, and A. Shrier. 1984. Global bifurcations of a periodically forced biological oscillator. *Phys. Rev. A* 29:1348-1357.

J. Gleick. 1987. *Chaos: Making a New Science.* Viking, New York.

A. L. Goldberger, V. Bhargava, B. J. West, and A. J. Mandell. 1985. On a mechanism of cardiac electrical stability. *Biophys. J.* 48:525-528.

J. Guckenheimer and P. Holmes. 1983. *Nonlinear Oscillations, Dynamical Systems, and Bifurcations of Vector Fields.* Springer-Verlag, New York.

H. Haken. 1983. *Synergetics: An Introduction,* 3rd. Ed. Springer-Verlag, New York.

B. Hess and M. Markus. 1987. Order and chaos in biochemistry. *Trends. Biochem. Sci.* 12:45-48.

A. V. Holden, ed. 1986. *Chaos.* Princeton University Press, Princeton, NJ.

B. Hoop, H. Kazemi, and L. S. Liebovitch. 1993. Rescaled range analysis of resting respiration. *Chaos* 3:27-29.

X.-J. Hou, R. Gilmore, G. B. Mindlin, and H. G. Solari. 1990. An efficient algorithm for fast O(N*ln(N)) box counting. *Phys. Lett.* A151:43-46.

P. M. Iannaccone and M. Khokha, eds. 1996. *Fractal Geometry in Biological*

Systems. CRC Press, Boca Raton, FL.

K. Kaneko. 1993. Chaotic traveling waves in a coupled map lattice. *Physica* D68:299-317.

D. T. Kaplan and R. J. Cohen. 1990. Is fibrillation chaos? *Circ. Res.* 67:886-892.

D. T. Kaplan and L. Glass. 1992. Direct test for determinism in a time series. *Phys. Rev. Lett.* 68:427-430.

J. A. S. Kelso. 1995. *Dynamical Patterns.* MIT Press, Cambridge, MA.

D. E. Lea and C. A. Coulson. 1949. The distribution of the number of mutants in bacterial populations. *J. Genetics* 49:264-285.

B. R. Levin, D. M. Gordon, and F. M. Stewart. 1989. Is natural selection the composer as well as the editor of genetic variation? Preprint.

L. S. Liebovitch. 1990. Fractal activity in cell membrane ion channels. In *Mathematical Approaches to Cardiac Arrhythmias*, ed. J. Jalife. *Ann. N.Y. Acad. Sci.* 591:375-391.

L. S. Liebovitch, J. Fischbarg, and J. P. Koniarek. 1987. Ion channel kinetics: A model based on fractal scaling rather than multistate Markov processes. *Math. Biosci.* 84:37-68.

L. S. Liebovitch and T. I. Tóth. 1989. A fast algorithm to determine fractal dimensions by box counting. *Phys. Lett.* A141:386-390.

L. S. Liebovitch and T. I. Tóth. 1991. A model of ion channel kinetics using deterministic chaotic rather than stochastic processes. *J. Theor. Biol.* 148:243-267.

E. N. Lorenz. 1963. Deterministic nonperiodic flow. *J. Atmos. Sci.* 20:130-141.

S. E. Luria and M. Delbruck. 1943. Mutations of bacteria from virus sensitivity of virus resistance. *Genetics* 28:491-511.

M. A. Mainster. 1990. The fractal properties of retinal vessels: Embryological and clinical implications. *Eye* 4:235-241.

B. B. Mandelbrot. 1974. A population birth-and-mutation process I: Explicit distributions for the number of mutants in an old culture of bacteria. *J. Appl. Prob.* 11:437-444.

B. B. Mandelbrot. 1983. *The Fractal Geometry of Nature.* W. H. Freeman and Company, New York.

M. Markus, D. Kuschmitz, and B. Hess. 1985. Properties of strange attractors in yeast glycolysis. *Biophys. Chem.* 22:95-105.

G. Mayer-Kress and S. P. Layne. 1987. Dimensionality of the human electroencephalogram. In *Perspectives in Biological Dynamics and Theoretical Medicine*, S. H. Koslow, A. J. Mandel, and M. F. Shlesinger, eds. *Ann. N.Y. Acad. Sci.* 504:62-87.

P. Meakin. 1986. Computer simulation of growth and aggregation processes. In *On Growth and Form: Fractal and Non-Fractal Patterns in Physics*, H. E. Stanley and N. Ostrowsky, eds. Martinus Nijhoff, Boston, pp. 111-135.

F. C. Moon. 1992. *Chaotic and Fractal Dynamics.* John Wiley & Sons, New York.

253

References

E. Mosekilde and L. Mosekilde, eds. 1991. *Complexity, Chaos, and Biological Evolution*. Plenum, New York.

M. A. H. Nerenberg and C. Essex. 1990. Correlation dimension and systematic geometric effects. *Phys. Rev. A* 42:7065-7074.

L. F. Olsen and W. M. Schaffer. 1990. Chaos versus noisy periodicity: Alternative hypotheses for childhood epidemics. *Science* 249:499-504.

A. R. Osborne and A. Provenzale. 1989. Finite correlation dimension for stochastic systems with power-law spectra. *Physica* D35:357-381.

E. Ott. 1993. *Chaos in Dynamical Systems*. Cambridge University Press, New York.

D. Paumgartner, G. Losa, and E. R. Weibel. 1981. Resolution effect on the stereological estimation of surface and volume and its interpretation in terms of fractal dimensions. *J. Micros.* 121:51-63.

C.-K. Peng, S. V. Buldyrev, A. L. Goldberger, S. Havlin, F. Sciortino, M. Simons, and H. E. Stanley. 1992. Long-range correlations in nucleotide sequences. *Nature* 356:168-170.

P. E. Rapp, T. R. Bashore, J. M. Martinerie, A. M. Albano, I. D. Zimmerman, and A. I. Mees. 1989. Dynamics of brain electrical activity. *Brain Tomography* 2:99-118.

L. F. Richardson. 1961. The problem of contiguity: An appendix to statistics of deadly quarrels. *General Systems Yearbook* 6:139-187.

R. Roy, T. W. Murphy Jr., T. D. Maier, and Z. Gills. 1992. Dynamical control of a chaotic laser: Experimental stabilization of a globally coupled system. *Phys. Rev. Lett.* 68:1259-1262.

R. H. Scanlan and J. W. Vellozzi. 1980. Catastrophic and annoying responses of long-span bridges to wind action. In *Long Span Bridges*, E. Cohen and B. Birdsall, eds. *Ann. N.Y. Acad. Sci.* 352:247-263.

W. M. Schaffer and M. Kot. 1986. Differential systems in ecology and epidemiology. In *Chaos*, A. V. Holden, ed. Princeton University Press, Princeton, NJ.

C. A. Skarda and W. J. Freeman. 1987. How brains make chaos in order to make sense out of the world. *Behav. Brain Sci.* 10:161-195.

L. A. Smith. 1988. Intrinsic limits of dimension calculations. *Phys. Lett.* A133:283-288.

T. G. Smith Jr., W. B. Marks, G. D. Lange, W. H. Sheriff, Jr., and E. A. Neale. 1988. A fractal analysis of cell images. *J. Neurosci. Meth.* 27:173-180.

F. Takens. 1981. Detecting strange attractors in turbulence. In *Dynamical Systems and Turbulence*, D. A. Rand and L.-S. Young, eds. Springer-Verlag, New York, pp. 366-381.

M. C. Teich. 1989. Fractal character of the auditory neural spike train. *IEEE Trans. Biomed. Engr.* 46:41-52.

M. C. Teich, D. H. Johnson, A. R. Kumar, and R. G. Turcott. 1990. Rate fluctuations and fractional power-law noise recorded from cells in the lower auditory pathway of the cat. *Hear. Res.* 46:41-52.

M. C. Teich, S. M. Khana, and S. E. Keilson. 1989. Nonlinear dynamics of cellular vibrations in the organ of Corti. *Acta Otolaryngol (Stockholm)* Suppl. 467:265-279.

J. Theiler, S. Eubank, A. Longtin, B. Garldrikian, and J. D. Farmer. 1992. Testing for nonlinearity in time series: The method of surrogate data. *Physica* D58:77-94.

B. J. West and A. L. Goldberger. 1987. Physiology in fractal dimensions. *Am. Sci.* 75:354-365.

S. Wolfram. 1994. *Cellular Automata and Complexity.* Addison-Wesley, Reading, MA.

A. Wolf, J. B. Swift, H. L. Swinney, and J. A. Vastano. 1985. Determining Lyapunov exponents from a time series. *Physica* D16:285-317.

N. Xu and J. Xu. 1988. The fractal dimension of the EEG as a physical measure of conscious brain activities. *Bull. Math. Biol.* 50:559-565.

J. P. Zbilut, G. Mayer-Kress, P. A. Sobotka, M. O'Toole, and J. X. Thomas, Jr. 1989. Bifurcations and intrinsic chaos and 1/f dynamics in an isolated perfused rat heart. *Biol. Cybern.* 61:371-381.

Illustration Credits

Figures reproduced with permission appear in the following graphics:

1.2.2	Caserta et al., 1990, Figs. 1(a), 1(b), 2(a).
1.2.2	West and Goldberger, 1987, Fig. 1.
1.2.3	Giaever and Keese, 1989, Figs. 1, 3.
1.2.3	Churilla et al., 1996, Fig. 1(c).
1.2.4	Liebovitch, 1990, Fig. 1.
1.3.3.	Richardson, 1961, Fig. 17.
1.3.4	Paumgartner et al., 1981, Fig. 9.
1.5.4	Meakin, 1986, Figs. 10, 12.
1.5.5	Brown and Scholz, 1985, Fig. 4.
1.5.6	Teich et al., 1990, Fig. 2(a).
1.5.7	Teich, 1989, Figs. 3(a), 3(b), 4(a), 5(a), 5(b).
1.5.8	Bassingwaighte et al., 1988, Fig. 4.
1.5.9	Hoop et al., 1993, Fig. 1.
1.5.12	Goldberger et al., 1985, Fig. 2.
1.5.12	Cargill et al., 1988, Fig 2(a).
2.2.6	Teich et al., 1989, Figs. 1,3.
2.2.8	Glass et al., 1984, Fig. 8(a), 8(b), 8(d), 8(f), 9(b).
2.2.9	Glass et al., 1984, Fig. 11(b1), 11(b2).
2.4.4	Hess and Markus, 1987, Fig. 2.
2.4.5	Markus et al., 1985, Fig. 1(b).
2.4.6	Kelso, 1995, Figs. 2.3, 2.4(a), 2.4(b), 2.6.
2.5.2	Schaffer and Kot, 1986, Fig. 8.5(a), 8.6(a), 8.8(a), 8.8(d).
2.5.4	Mayer-Kress and Layne, 1987, Fig. 3.
2.6.2	Roy et al., 1992, Figs. 2(a), 2(b), 2(d).
3.1.2	Bak et al., 1988, Fig. 1.

Index

Index